Reinforced Concrete Beams, Columns and Frames

Reinforced Concrete Beams, Columns and Frames

Section and Slender Member Analysis

Jostein Hellesland
Noël Challamel
Charles Casandjian
Christophe Lanos

First published 2013 in Great Britain and the United States by ISTE Ltd and John Wiley & Sons, Inc.

ISTE Ltd
27-37 St George's Road
London SW19 4EU
UK

www.iste.co.uk

John Wiley & Sons, Inc.
111 River Street
Hoboken, NJ 07030
USA

www.wiley.com

Library of Congress Control Number: 2012954708

British Library Cataloguing-in-Publication Data
A CIP record for this book is available from the British Library
ISBN: 978-1-84821-569-6

Printed and bound in Great Britain by CPI Group (UK) Ltd., Croydon, Surrey CR0 4YY

Table of Contents

Preface

The authors have written two books on the theoretical and practical design of reinforced concrete beams, columns and frame structures. The book entitled *Reinforced Concrete Beams, Columns and Frames – Mechanics and Design* deals with the fundamental aspects of the mechanics and design of reinforced concrete in general, related to both the *Serviceability Limit State (SLS)* and the *Ultimate Limit State (ULS)*. This book, entitled *Reinforced Concrete Beams, Columns and Frames – Section and Slender Member Analysis*, deals with more advanced *ULS* aspects, along with instability and second-order analysis aspects. The two books are complementary, and, indeed, could have been presented together in one book. However, for practical reasons, it has proved more convenient to present the material in two separate books with the same preface in both books.

The books are based on an analytical approach for designing usual reinforced concrete structural elements, compatible with most international design rules, including for instance the European design rules *Eurocode 2* for reinforced concrete structures. The presentations have tried to distinguish between what belongs to the philosophy of structural design of such structural elements (related to strength of materials arguments) and the design rules aspects associated with specific characteristic data (for the material or the loading parameters). The *Eurocode 2* design rules are used in most of the examples of applications in the books. Even so, older international rules, as well as national rules such as the old French rules *BAEL* ("Béton Armé aux Etats Limites", or Reinforced Concrete Limit State in English) will sometimes be mentioned, at least for historical reasons.

Whatever the design rules considered, the fundamental concept of *Limit State* will be detailed, and more specifically, the *Serviceability Limit State (SLS)* and *Ultimate Limit State (ULS)* both in bending and in compression will be investigated.

The books are devoted mainly to the bending (flexural) behavior of reinforced concrete elements, including geometrical nonlinear effects (in this book). However, two major aspects of reinforced concrete design are not treated. These are shear force effects and the calculation of crack width as dealt with in the *Crack Opening Limit State* in *Eurocode 2*. The latter represents a major new contribution as compared to some older European rules such as *BAEL*. The readers are referred to the very good monographs devoted to the general presentation of *Eurocode 2* for these additional parts (see for instance [CAL 05]; [DES 05]; [MOS 07]; [EUR 08]; [PAI 09]; [PER 09]; [ROU 09a]; [ROU 09b]; [THO 09]; [PER 10]; [SIE 10]; [PAU 11]).

We would also like to point out that the calculation of crack widths, even under a simple loading configuration, such as uniform tension loading, still remains a difficult topic. Besides, the authors are even convinced that meaningful efforts should be addressed in the future, for facilitating the transfer of knowledge from theoretical research in fracture or damage mechanics, to applied, practical design rules. In connection with this, cohesive crack models were introduced in the 1970s to investigate the crack opening in mode I of failure [HIL 76], whereas non-local damage mechanics models were developed in the 1980s for efficient computations of damage softening materials [PIJ 87]. Both appear to belong to the families of non-local models which contain an internal length, for the control of the post-failure process [PLA 93]. Non-local damage mechanics is now widely used in the research community for the study of reinforced concrete structures (see for instance [BAZ 03]; [MAZ 09]). The authors of these books have also conducted some research in this field to better understand the failure of some simple reinforced concrete structural elements (research at INSA, Rennes, University of Rennes I, University of South Brittany or University of Oslo – see for instance [CHA 05]; [CHA 06]; [CHA 07]; [CHA 08]; [CHA 09]; [CHA 10]; [CHA 11]; [CHA 12]). However, the engineering community has not yet necessarily integrated these results into the design process or even into the rules. The gap between the research activity and the engineering methodology is probably too large at present, and researchers will probably have some responsibility in the future to make their results more tractable to the engineering community. With respect to these books, some very simple concepts of non-local mechanics will be presented when necessary. However, the books are mainly devoted to the design of a reinforced concrete structure at a given limit state, the cracking evolution problem often being considered as a secondary problem. We have chosen to concentrate our efforts on the bending design based on the pivot concept, at both the *Serviceability Limit State (SLS)* and the *Ultimate Limit State (ULS)*. The last part of this book deals with the design of columns against buckling, and how to take into account second-order effects will be presented for stability design. In particular, some engineering approaches practiced by engineers will be detailed, to replace efficiently, when possible, the nonlinear evolution problem associated with micro-cracking and failure.

The books are aimed at both undergraduate and graduate (Licence and master) students in civil engineering, engineers and teachers in the field of reinforced concrete design. In addition, researchers and PhD students can find something of interest in the books, including the presentation on elementary applications of non-local damage or plasticity mechanics applied to the ultimate bending of reinforced concrete beams (and columns). We hope that the basic ideas presented in the books can contribute to stimulating the links between the research community in this field (computational modeling and structural analysis) and the design community with practical structural cases. The principles of limit state design will be introduced and developed first, both at both the Serviceability limit state (SLS) and the Ultimate Limit State (ULS), illustrated by some detailed examples to illustrate the introduced methodology.

Older books (see for instance [HOG 51]; [BAK 56]; [SAR 68]; [ROB 74]; [PAR 75]; [FUE 78]; [LEO 78]; [ALB 81]; [LEN 81]; [BAI 83]; [GYO 88]; [WAL 90]; [PAU 92]; [MAC 97]) have been used in some portions of the books (for establishing familiar and well known equations on section design (in particular equations based on the simplified rectangular stress-strain diagram for concrete in compression). In particular, the authors want to acknowledge the very exhaustive work of Professor Robinson, at Ecole Nationale des Ponts et Chaussées, whose reinforced concrete teaching book published in 1974 can still be considered as a main reference with modern insights into reinforced concrete design [ROB 74]. We have also been inspired by the more recent and very exhaustive works of Professor Thonier (see for instance [THO 09]), also at Ecole Nationale des Ponts et Chaussées.

The previous book, *Reinforced Concrete Beams, Columns and Frames –
Mechanics and Design*, is organized as follows. Chapters 1 and 2 deal with
the Serviceability Limit State, for both the design and the cross-section
verification. The French school of reinforced concrete design have
commonly used the concept of "Pivot", which is related to the limit behavior
of the cross-section with respect to the steel and concrete material
characteristics. The Pivot A (where the steel material characteristics control
the behavior of the cross-section at the Limit State), and Pivot B (where the
concrete material characteristics control the behavior of the cross-section at
the Limit State) concepts are introduced with the Serviceability Limit State
in Chapter 1. Chapter 1 is mainly focused on the design aspects, whereas
Chapter 2 deals with the verification of the reinforced concrete section with
both the bending and the normal forces effects. The general theory presented
in these first two chapters is valid for arbitrary shapes of the reinforced
concrete cross-section including for instance rectangular, triangular,
trapezoidal or T-cross-sections. Chapter 2 ends with the presentation of a
cubic equation for the determination of the neutral axis in the general
loading configuration, including the normal force effects. This elegant
equation is also known as the cubic equation of the French reinforced
concrete design rules dating from 1906 (*Circulaire du 20 Octobre 1906*)
(and reported in the book by Magny, [MAG 14]) or those dating from 1934
(*Règlements des marchés de l'état de 1934* – also in French), also recently
reported by Professor Thonier for T-cross-sections [THO 09]. Finally, the
tension stiffening phenomenon is introduced in terms of a nonlinear bending
moment-curvature constitutive law and some verification examples are given
to illustrate the theoretical results obtained in the fundamental parts.

Chapters 3 and 4 focus on the fundamental aspects of the Ultimate Limit
State. Chapter 3 starts with a brief introduction to the concept of the
Ultimate Limit State for the bending of a reinforced concrete beam. The
need to use some non-local theory to correctly model the post-failure
behavior of reinforced concrete structural elements is shown in the presence
of global curvature softening. The material characteristics of the steel and
concrete allowed by Eurocode 2 are listed, and compared with each other. It
is possible to derive analytically the normal forces and the resultant bending
moment in the compression block for each considered concrete law,
including the parabolic-rectangle constitutive law, the simplified rectangular
constitutive law, the bilinear constitutive law or Sargin's nonlinear
constitutive law. These preliminaries will be used later for the design of
reinforced concrete sections at Ultimate Limit State. Chapter 4 discussed

some possible bending moment – curvature law of typical reinforced concrete sections. These cross-sectional behaviors can be deduced from the local characteristics of the steel and concrete constituents. The relevancy of a bilinear approximation for the moment-curvature constitutive law is discussed, with possible tractable analytical results for engineering purposes. Chapter 4 concludes with some buckling and post-buckling results obtained for a reinforced concrete column modeled with a simplified nonlinear bending-curvature constitutive law. It is shown that reinforced concrete columns typically behave like imperfection-sensitive structural systems.

The current book, *Reinforced Concrete Beams, Columns and Frames – Section and Slender Member Analysis*, is organized as follows. The advanced design of general reinforced concrete sections is treated in Chapter 1. The reinforced concrete section can be optimized for a given loading (in term of minimization of the steel quantity for instance), with some constrained equations. Also discussed is how the Serviceability and the Ultimate Limit States can be compared, depending on the material and loading features of the problem. A design of the cross-section in biaxial bending is also proposed. More generally in this chapter, the reinforced concrete section is designed for various constitutive laws for concrete and the steel behavior, including possible steel hardening, with possible analytical solutions for the optimized design. Some design examples are included for the various solicitations including simple bending, bending combined with normal forces or bi-axial bending. The last part of Chapter 1 discusses the possible use of moment-normal forces interaction diagrams available in international codes, and some new possible improvements of these simplified diagrams.

Chapter 2 is devoted to general aspects of instability of and second-order effects in slender compression members, and in frames that include such members. For such cases, it is necessary to consider second-order load effects in the analysis and design. The concepts of braced, unbraced and partially braced systems as well as associated moment formulations are presented, and the useful distinction between local and global second-order effects discussed. The general principles of analysis and design of individual reinforced concrete columns and frame systems are reviewed in order to provide a general understanding of the problem area. This includes a presentation and discussion of fundamental concepts and theory behind approximate analysis and design methods to provide a reasonable complete basis for relevant analysis and design requirements as given in existing design rules, such as in Eurocode 2. This also includes a discussion of the applicability of equivalent elastic analysis as an approximation to nonlinear analyses (accounting for both material and geometric nonlinear effects). Local and global slenderness limits, allowing second-order effects to be neglected, are presented and discussed. Chapter 3 deals with approximate analysis methods used for efficient and practical elastic stability calculations, and second-order elastic sway and moment calculations. Included in this chapter are different methods for computing effective lengths, and methods employing the widely used effective length concept in frame analysis. Basic concepts are explained and simple and more complex engineering examples are included to provide a better understanding of the methods.

Reinforced Concrete Beams, Columns and Frames – Mechanics and Design along with the first chapter of the current book were mainly written by Charles Casandjian, Noël Challamel and Christophe Lanos, whereas Jostein Hellesland mostly contributed to the final two chapters of this book.

Finally, an appendix is provided that gives further developments on the theoretical background of Cardano's method, useful for the resolution of a cubic equation, often encountered in the designing of a reinforced concrete section at both Serviceability and Ultimate Limit States. An appendix giving a table of steel diameters is also provided for the quick and efficient selection of reinforcement sizes in design calculations.

Charles CASANDJIAN,
Noël CHALLAMEL,
Christophe LANOS and
Jostein HELLESLAND
December 2012

Chapter 1

Advanced Design at Ultimate Limit State (ULS)

1.1. Design at ULS – simplified analysis

1.1.1. *Simplified rectangular behavior – rectangular cross-section*

1.1.1.1. *Simplified rectangular behavior – rectangular cross-section with only tensile steel reinforcement*

In this section, the design of a reinforced concrete section at the ultimate limit state (ULS) is considered by using a rectangular simplified law for the compression concrete block, and a bilinear law for the steel that accounts for the hardening behavior. This design is compatible with Eurocode 2 material parameters. The rectangular cross-section is shown in Figure 1.1. The steel reinforcement area has to be designed for this given concrete section.

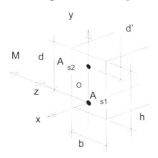

Figure 1.1. *Rectangular cross-section at ultimate limit state*

Pivot AB is characterized for the rectangular cross-section by the neutral axis position:

$$\alpha_{AB} = \frac{\varepsilon_{cu3}}{\varepsilon_{cu3} - \varepsilon_{ud}} \qquad [1.1]$$

For instance, for a C30-37 type concrete and for a B500B steel, the numerical values are $\varepsilon_{cu3} = -3.5\%_0$ and $\varepsilon_{ud} = 0.9\varepsilon_{uk} = 45\%_0$ (if the hardening effect is taken into account), leading to $\alpha_{AB} = 0.072165$.

For a general reinforced concrete section, the bending moment and normal force equilibrium equations are written with respect to the center of gravity of the tensile steel reinforcement as:

$$\begin{cases} M_u = M_c - A_{s2}\sigma_{s2}(d - d') \\ N_u = N_c + \sigma_{s1}A_{s1} + \sigma_{s2}A_{s2} \end{cases} \text{with } y = \alpha d \qquad [1.2]$$

where N_c and M_c are the normal force and moment in the compression concrete block calculated from the simplified rectangular constitutive law.

$$\begin{cases} M_c = -0.8yf_{cu}b(d - 0.4y) \\ N_c = bf_{cu}0.8y \end{cases} \qquad [1.3]$$

For a reinforced concrete section with only steel reinforcement, the bending moment equilibrium equation is written in a dimensionless format:

$$\mu = \mu_u = 0.8\alpha(1 - 0.4\alpha) \text{ with } \mu = \frac{-M_c}{bd^2 f_{cu}} \quad \text{and} \quad \mu_u = \frac{-M_u}{bd^2 f_{cu}} \qquad [1.4]$$

We recognize a second-order equation with respect to the position of the neutral axis α:

$$\alpha^2 - \frac{5}{2}\alpha + \frac{25}{8}\mu_u = 0 \qquad [1.5]$$

whose solution of interest is given by:

$$\alpha = 1.25(1 - \sqrt{1 - 2\mu})$$ [1.6]

Note that this equation is independent of the pivot considered (pivot A or pivot B). Once the position of the neutral axis is calculated, the tensile steel area is obtained from the normal force equilibrium equation:

$$A_{s1} = \frac{N_u - bd\psi\alpha f_{cu}}{\sigma_{s1}} \text{ with } \psi = 0.8$$ [1.7]

$\psi = 0.8$ for the rectangular simplified constitutive law for concrete.

In the case of pivot A, for $\alpha \le \alpha_{AB}$, the strain capacity of the tensile steel reinforcement ε_{s1} is equal to ε_{ud}, and the steel stress σ_{s1} is equal to $q\varepsilon_{ud} + q'$ (see equation [3.91] of [CAS 12]), leading to:

$$\alpha \le \alpha_{AB} \Rightarrow A_{s1} = \frac{N_u - bd\psi\alpha f_{cu}}{q\varepsilon_{ud} + q'}$$ [1.8]

In the case of pivot B, for $\alpha \le \alpha_{AB}$, the strain of the tensile steel reinforcement ε_{s1} depends on the position of the neutral axis. If the tensile steel reinforcement behaves in elasticity, the tensile steel area is calculated from:

$$\alpha \ge \alpha_{AB} \Rightarrow A_{s1} = \frac{N_u - bd\psi\alpha f_{cu}}{E_s \frac{\alpha - 1}{\alpha}\varepsilon_{cu3}} \text{ if } \varepsilon_{s1} = \frac{\alpha - 1}{\alpha}\varepsilon_{cu3} \le \varepsilon_{su} = \frac{f_{su}}{E_s}$$ [1.9]

In pivot B, the tensile steel reinforcement behaves in the elastic range for:

$$\varepsilon_{s1} = \frac{\alpha - 1}{\alpha}\varepsilon_{cu3} \le \varepsilon_{su} = \frac{f_{su}}{E_s} \Leftrightarrow \alpha \ge \alpha_{bal} = \frac{-\varepsilon_{cu3}}{\varepsilon_{su} - \varepsilon_{cu3}}$$ [1.10]

In pivot B, the behavior of the reinforced cross-section, when the tensile steel reinforcement reaches the elastic strain limit $\varepsilon_{s1} = \varepsilon_{su}$ is called the balance failure behavior $\alpha = \alpha_{bal}$. This behavior is observed in the presence of compression normal forces. However, in simple bending (without normal forces), the design of the reinforced cross-section when the steel reinforcements behave linearly elastically may not be efficient for economic reasons, as the tensile steel reinforcements in the elasticity range are not optimized. In this case, it is recommended to add some compression steel reinforcement to increase the stress level in the compression steel reinforcement.

If the tensile steel reinforcement behaves in plasticity, the tensile steel area has to be calculated from:

$$\alpha \geq \alpha_{AB} \Rightarrow A_{s1} = \frac{N_u - bd\psi\alpha f_{cu}}{q\frac{\alpha-1}{\alpha}\varepsilon_{cu3} + q'} \text{ if } \varepsilon_{s1} = \frac{\alpha-1}{\alpha}\varepsilon_{cu3} \in [\varepsilon_{su};\varepsilon_{ud}] \quad [1.11]$$

Consider, for instance, the case of the reinforced concrete section in simple bending ($N_u = 0$). The tensile steel reinforcement is assumed to be perfectly plastic without hardening ($k = 1$ and then $q = 0$ and $q' = f_{su}$). For the perfect plasticity case considered in these numerical applications, pivot A does not exist as there is no strain limit capacity imposed by the Eurocode 2 rules (or the steel ductility is so high that it has no limited effect in the design). The steel content is obtained with the simplified rectangular diagram for the concrete constitutive law as:

$$\alpha \leq \alpha_{bal} \Rightarrow \omega_1 = \alpha\psi = 1 - \sqrt{1 - 2\mu}; \alpha \geq \alpha_{bal}$$

$$\Rightarrow \omega = \psi\frac{\alpha^2}{1-\alpha}\left(\frac{\varepsilon_{su}}{-\varepsilon_{cu3}}\right) \text{ with } \alpha_{bal} = \frac{-\varepsilon_{cu3}}{\varepsilon_{su} - \varepsilon_{cu3}} \text{ and}$$

$$\omega_1 = \frac{A_{s1}f_{su}}{-f_{cu}bd} = \rho'_{s1}\left(\frac{f_{su}}{-f_{cu}}\right) \quad [1.12]$$

For a C30-37 type concrete and for a B500B steel, the numerical values of the ultimate limit strains, are $\varepsilon_{cu3} = -3.5‰$ and $\varepsilon_{su} = 2.17‰$, leading to $\alpha_{bal} = 0.617$ and $\mu_{bal} = 0.372$.

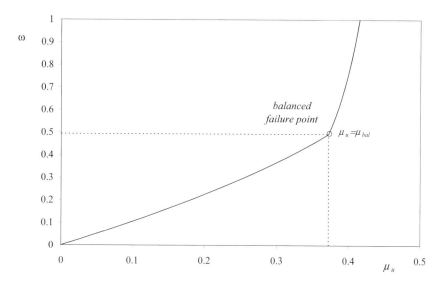

Figure 1.2. *Evolution of the normalized steel content ω with respect to the reduced moment μ_u for a reinforced concrete section without compression steel reinforcement;*

$$\omega = \frac{A_s f_{su}}{-f_{cu} bd} \; ; \; A_{s1} = A_s \; ; \; pivot \; B$$

Figure 1.2 shows the change of slope of the normalized steel content ω with respect to the reduced moment μ_u, at the balanced failure point, for a C30-37 type concrete and for a B500B steel. This change of slope is not related to the transition between pivot A and pivot B, as observed at the serviceability limit state (SLS), but is clearly due to the transition toward the balance failure point (transition from plasticity to elasticity in the tensile steel reinforcement) at pivot B ULS (see also Figure 4.10 based on a bilinear constitutive law for the compression concrete block).

1.1.1.2. Simplified rectangular behavior – rectangular cross-section with both tensile and compression steel reinforcement

It has been shown that the determination of the neutral axis position α at the ULS, with a simplified rectangular constitutive law for the concrete behavior, does not depend on the pivot nature (pivot A or pivot B) for a reinforced concrete section with only some tensile steel reinforcement ($A_{s2} = 0$). This fundamental property is no more valid in presence of both tensile and compression steel reinforcements.

1.1.1.2.1. Simplified rectangular behavior – pivot A

The case of pivot A is first analyzed, characterized by the strain values:

$$\alpha \le \alpha_{AB} \Rightarrow \varepsilon_{s2} = -\varepsilon_{ud} \frac{\alpha - \delta'}{1 - \alpha} \text{ and } \sigma_{s2} = m\varepsilon_{s2} + m' \qquad [1.13]$$

where m and m' are the parameters of the hardening plasticity constitutive law, defined for instance by equation [3.90] [CAS 12], that depend on the strain value of the compression steel reinforcement. As already introduced, δ' is the geometrical parameter defined by $\delta' = d'/d$. The bending moment equilibrium equation is then written as:

$$M_u = -0.8\alpha bd^2 f_{cu}(1 - 0.4\alpha) - A_{s2}d(1 - \delta')\left[-m\varepsilon_{ud} \frac{\alpha - \delta'}{1 - \alpha} + m' \right] \quad [1.14]$$

The dimensionless equation is then obtained as:

$$\mu_u = 0.8\alpha(1 - 0.4\alpha) + \rho'_{s2}(1 - \delta')\left[\frac{-m}{f_{cu}}\varepsilon_{ud} \frac{\alpha - \delta'}{1 - \alpha} + \frac{m'}{f_{cu}} \right] \qquad [1.15]$$

where the normalized compression steel quantity is calculated as $\rho'_{s2} = A_{s2}/bd$. The determination of the dimensionless position of neutral axis α is then obtained from the resolution of a cubic equation:

$$\alpha^3 - \frac{7}{2}\alpha^2 + \alpha\left[\frac{25}{8}\mu_u + \frac{5}{2} + \frac{25}{8}\frac{\rho'_{s2}(1 - \delta')}{(-f_{cu})}(m\varepsilon_{ud} + m') \right]$$

$$- \left[\frac{25}{8}\mu_u + \frac{25}{8}\frac{\rho'_{s2}(1 - \delta')}{(-f_{cu})}(m\varepsilon_{ud}\delta' + m') \right] = 0 \qquad [1.16]$$

Equation [1.16] can be solved using Cardano's method (see Appendix 1 for the application of Cardano's method). The values of m and m' depend on the strain level in the compression steel reinforcement. The characteristic value associated with the boundary between the elastic range and the plastic range of the compression steel reinforcement is given in pivot A by:

$$\varepsilon_{s2} = -\varepsilon_{su} = -\varepsilon_{ud} \frac{\alpha - \delta'}{1 - \alpha} \Rightarrow \alpha = \frac{\varepsilon_{su} + \delta'\varepsilon_{ud}}{\varepsilon_{su} + \varepsilon_{ud}} \qquad [1.17]$$

Therefore, two stages can be defined in pivot A depending on the strain level in the compression steel reinforcement. The compression steel reinforcement behaves in the elastic range for:

$$\alpha \in \left[0; \inf\left(\frac{-\varepsilon_{cu3}}{\varepsilon_{ud} - \varepsilon_{cu3}}; \frac{\varepsilon_{su} + \delta'\varepsilon_{ud}}{\varepsilon_{su} + \varepsilon_{ud}} \right) \right] \Rightarrow m = E_s \text{ and } m' = 0 \qquad [1.18]$$

The compression steel reinforcement behaves in the plastic range for:

$$\alpha \in \left[\inf\left(\frac{-\varepsilon_{cu3}}{\varepsilon_{ud} - \varepsilon_{cu3}}; \frac{\varepsilon_{su} + \delta'\varepsilon_{ud}}{\varepsilon_{su} + \varepsilon_{ud}} \right); \frac{-\varepsilon_{cu3}}{\varepsilon_{ud} - \varepsilon_{cu3}} \right] \Rightarrow m = q \text{ and } m' = -q' \text{ [1.19]}$$

The hardening parameters q and q' are defined by equation [3.91] [CAS 12]. In the particular case of unsymmetrical reinforced concrete section without compression steel reinforcement, the cubic equation is simplified in:

$$\rho'_{s2} = 0 \Rightarrow \alpha^3 - \frac{7}{2}\alpha^2 + \left(\frac{25}{8}\mu_u + \frac{5}{2} \right)\alpha - \frac{25}{8}\mu_u = 0 \qquad [1.20]$$

Equation [1.20] can be easily factorized by:

$$\rho'_{s2} = 0 \Rightarrow (\alpha - 1)\left(\alpha^2 - \frac{5}{2}\alpha + \frac{25}{8}\mu_u \right) = 0 \qquad [1.21]$$

We recognize the second-order equation already introduced in equation [1.5] for the rectangular reinforced cross-section with only tensile steel reinforcement.

In the case of pivot A, for $\alpha \leq \alpha_{AB}$, the tensile steel area is computed from the normal force equilibrium equation written as:

$$\alpha \leq \alpha_{AB} \Rightarrow A_{s1} = \frac{N_u - bd\psi\alpha f_{cu} - A_{s2}\left(-m\varepsilon_{ud}\dfrac{\alpha - \delta'}{1 - \alpha} + m' \right)}{q\varepsilon_{ud} + q'} \qquad [1.22]$$

where the values of m and m' in the compression steel reinforcement depend on the α value according to equations [1.18] and [1.19].

1.1.1.2.2. Simplified rectangular behavior – pivot B

The case of pivot B is now analyzed, characterized by the strain values:

$$\alpha \geq \alpha_{AB} \Rightarrow \varepsilon_{s2} = \varepsilon_{cu3}\frac{\alpha - \delta'}{\alpha} \text{ and } \sigma_{s2} = m\varepsilon_{s2} + m' \qquad [1.23]$$

where m and m' are the parameters of the hardening plasticity constitutive law, defined for instance by equation [3.90] [CAS 12], that depend on the strain value of the compression steel reinforcement. The bending moment equilibrium equation is written in dimensionless format as:

$$\mu_u = 0.8\alpha(1 - 0.4\alpha) + \rho'_{s2}(1 - \delta')\left[\frac{m}{f_{cu}}\varepsilon_{cu3}\frac{\alpha - \delta'}{\alpha} + \frac{m'}{f_{cu}}\right] \qquad [1.24]$$

The determination of the dimensionless position of neutral axis α is then obtained from the resolution of a cubic equation:

$$\alpha^3 - \frac{5}{2}\alpha^2 + \alpha\left[\frac{25}{8}\mu_u + \frac{25}{8}\frac{\rho'_{s2}(1 - \delta')}{(-f_{cu})}(m\varepsilon_{cu3} + m')\right]$$

$$-\left[\frac{25}{8}\frac{\rho'_{s2}(1 - \delta')}{(-f_{cu})}m\varepsilon_{cu3}\delta'\right] = 0 \qquad [1.25]$$

Equation [1.25] can also be solved using Cardano's method (see Appendix 1 for the application of Cardano's method). The values of m and m' depend on the strain level in the compression steel reinforcement. The characteristic value associated with the boundary between the elastic range and the plastic range of the compression steel reinforcement is given in pivot B, by:

$$\varepsilon_{s2} = -\varepsilon_{su} = \varepsilon_{cu3}\frac{\alpha - \delta'}{\alpha} \Rightarrow \alpha = \frac{\delta'\varepsilon_{cu3}}{\varepsilon_{su} + \varepsilon_{cu3}} \qquad [1.26]$$

Therefore, two stages can be defined in pivot B depending on the strain level in the compression steel reinforcement. The compression steel reinforcement behaves in the elastic range for:

$$\alpha \in \left[\frac{-\varepsilon_{cu3}}{\varepsilon_{ud} - \varepsilon_{cu3}}; \sup\left(\frac{-\varepsilon_{cu3}}{\varepsilon_{ud} - \varepsilon_{cu3}}; \frac{\delta' \varepsilon_{cu3}}{\varepsilon_{su} + \varepsilon_{cu3}} \right) \right] \Rightarrow m = E_s \text{ and } m' = 0 \qquad [1.27]$$

The compression steel reinforcement behaves in the plastic range for:

$$\alpha \in \left[\sup\left(\frac{-\varepsilon_{cu3}}{\varepsilon_{ud} - \varepsilon_{cu3}}; \frac{\delta' \varepsilon_{cu3}}{\varepsilon_{su} + \varepsilon_{cu3}} \right); \frac{h}{d} \right] \Rightarrow m = q \text{ and } m' = -q' \qquad [1.28]$$

The hardening parameters q and q' are defined by equation [3.91] [CAS 12]. In the particular case of a unsymmetrical reinforced concrete section without compression steel reinforcement, the cubic equation is a mathematically degenerate second-order equation, simplified in:

$$\rho'_{s2} = 0 \Rightarrow \alpha^2 - \frac{5}{2}\alpha + \frac{25}{8}\mu_u = 0 \qquad [1.29]$$

In the case of pivot B, for $\alpha \geq \alpha_{AB}$, the strain of the tensile steel reinforcement ε_{s1} depends on the position of the neutral axis. If the tensile steel reinforcement behaves in elasticity, the tensile steel area is calculated from:

$$\alpha \geq \alpha_{AB} \Rightarrow A_{s1} = \frac{N_u - bd\psi\alpha f_{cu} - A_{s2}\left(m\varepsilon_{cu3}\dfrac{\alpha - \delta'}{\alpha} + m' \right)}{E_s \dfrac{\alpha - 1}{\alpha}\varepsilon_{cu3}}$$

$$\text{if } \varepsilon_{s1} = \frac{\alpha - 1}{\alpha}\varepsilon_{cu3} \leq \varepsilon_{su} = \frac{f_{su}}{E_s} \qquad [1.30]$$

If the tensile steel reinforcement behaves in plasticity, the tensile steel area has to be calculated from:

$$\alpha \geq \alpha_{AB} \Rightarrow A_{s1} = \frac{N_u - bd\psi\alpha f_{cu} - A_{s2}\left(m\varepsilon_{cu3}\dfrac{\alpha - \delta'}{\alpha} + m' \right)}{q\dfrac{\alpha - 1}{\alpha}\varepsilon_{cu3} + q'}$$

$$\text{if } \varepsilon_{s1} = \frac{\alpha - 1}{\alpha}\varepsilon_{cu3} \in [\varepsilon_{su}; \varepsilon_{ud}] \qquad [1.31]$$

where the values of m and m' in the compression steel reinforcement depend on the α value according to equations [1.27] and [1.28].

1.1.1.3. Optimization of the steel reinforcement

In this section, the optimization problem of finding the strongest reinforced concrete section for a given amount of steel density is studied. The section is reinforced by both some tensile and some compression steel reinforcement. The dimensionless steel ratio is introduced as:

$$\omega_1 = \frac{A_{s1}f_{su}}{-bdf_{cu}}, \omega_2 = \frac{A_{s2}f_{su}}{-bdf_{cu}}, \omega = \omega_1 + \omega_2 \text{ and } A_s = A_{s1} + A_{s2} \qquad [1.32]$$

The section will be optimized with respect to the steel area ratio, which can be expressed by a rotation parameter θ as:

$$\omega_1 = \omega\cos^2\theta, \omega_2 = \omega\sin^2\theta \text{ and then } \theta = \arctan\sqrt{\frac{\omega_2}{\omega_1}} \qquad [1.33]$$

The concrete is modeled with the simplified rectangular constitutive law, and the steel reinforcements are modeled with the elastic and perfectly plastic constitutive law. It means that the section is controlled by pivot B. For most applications, the compression steel reinforcements are in the plasticity range when the section reaches the balance failure point, which is mathematically expressed by the inequality:

$$\frac{\delta'\varepsilon_{cu3}}{\varepsilon_{su} + \varepsilon_{cu3}} \leq \alpha_{bal} = \frac{-\varepsilon_{cu3}}{\varepsilon_{su} - \varepsilon_{cu3}} \qquad [1.34]$$

The position of neutral axis for a given steel quantity is determined from the normal force equilibrium equation, which is written when the tensile reinforcement works in plasticity and the compression reinforcement in elasticity as:

$$\frac{\delta'\varepsilon_{cu3}}{\varepsilon_{su} + \varepsilon_{cu3}} \leq \alpha \Rightarrow A_{s1}f_{su} = -bd\psi\alpha f_{cu} - A_{s2}E_s\varepsilon_{cu3}\frac{\alpha - \delta'}{\alpha} \qquad [1.35]$$

where $\psi = 0.8$, as usual for the rectangular simplified constitutive law used in the compression concrete. This equation leads to a second-order equation for the calculation of the position of neutral axis as:

$$\alpha^2 - \frac{\omega\alpha}{\psi}\left(\cos^2\theta + \frac{\varepsilon_{cu3}}{\varepsilon_{su}}\sin^2\theta\right) + \frac{\omega}{\psi}\delta'\frac{\varepsilon_{cu3}}{\varepsilon_{su}}\sin^2\theta = 0 \qquad [1.36]$$

whose solution of interest is represented by:

$$\alpha = \frac{\omega}{2\psi}\left(\cos^2\theta + \frac{\varepsilon_{cu3}}{\varepsilon_{su}}\sin^2\theta\right)$$
$$+ \frac{1}{2}\sqrt{\frac{\omega^2}{\psi^2}\left(\cos^2\theta + \frac{\varepsilon_{cu3}}{\varepsilon_{su}}\sin^2\theta\right)^2 - \frac{4\omega}{\psi}\delta'\frac{\varepsilon_{cu3}}{\varepsilon_{su}}\sin^2\theta} \qquad [1.37]$$

The reduced moment in pivot B with the compression steel reinforcement in the elastic range is given by:

$$\mu_u = 0.8\alpha(1 - 0.4\alpha) + \omega(\sin^2\theta)\frac{(1 - \delta')(\alpha - \delta')}{\alpha}\frac{(-\varepsilon_{cu3})}{\varepsilon_{su}} \qquad [1.38]$$

When both the tensile and the compression reinforcements work in plasticity, the normal force equilibrium equation gives the neutral axis position from:

$$\frac{\delta'\varepsilon_{cu3}}{\varepsilon_{su} + \varepsilon_{cu3}} \le \alpha \le \frac{-\varepsilon_{cu3}}{\varepsilon_{su} - \varepsilon_{cu3}} \Rightarrow A_{s1}f_{su} = -bd\psi\alpha f_{cu} + A_{s2}f_{su} \qquad [1.39]$$

The neutral axis position is then directly obtained from a linear relationship:

$$\alpha = \frac{\omega\cos(2\theta)}{\psi} \qquad [1.40]$$

The reduced moment in pivot B with the compression steel reinforcement in the plastic range is given by:

$$\mu_u = 0.8\alpha(1-0.4\alpha) + \omega(\sin^2\theta)(1-\delta') \qquad [1.41]$$

The last case is the case where the compression steel reinforcements work in plasticity whereas the tensile steel reinforcements are now in elasticity. This transition, also called the balanced failure mode, is characterized by the following normal force equilibrium equation:

$$\frac{-\varepsilon_{cu3}}{\varepsilon_{su}-\varepsilon_{cu3}} \leq \alpha \leq \frac{h}{d} \Rightarrow A_{s1}f_{su}\frac{1-\alpha}{\alpha}\left(\frac{-\varepsilon_{cu3}}{\varepsilon_{su}}\right) = -bd\psi\alpha f_{cu} + A_{s2}f_{su} \qquad [1.42]$$

We also obtain a second-order equation for the determination of the neutral axis position:

$$\alpha^2 + \frac{\omega}{\psi}\left[\sin^2\theta + \left(\frac{-\varepsilon_{cu3}}{\varepsilon_{su}}\right)\cos^2\theta\right]\alpha - \frac{\omega}{\psi}\cos^2\theta\left(\frac{-\varepsilon_{cu3}}{\varepsilon_{su}}\right) = 0 \qquad [1.43]$$

whose solution of interest is given by:

$$\alpha = \frac{-\omega}{2\psi}\left(\sin^2\theta - \frac{\varepsilon_{cu3}}{\varepsilon_{su}}\cos^2\theta\right)$$
$$+ \frac{1}{2}\sqrt{\frac{\omega^2}{\psi^2}\left(\sin^2\theta - \frac{\varepsilon_{cu3}}{\varepsilon_{su}}\cos^2\theta\right)^2 - 4\frac{\omega\,\varepsilon_{cu3}}{\psi\,\varepsilon_{su}}\cos^2\theta} \qquad [1.44]$$

Equation [1.41] is still valid for the reduced moment in the range of parameters considered.

For a C30-37 type concrete and for a B500B steel, the numerical values of the ultimate limit strains, are $\varepsilon_{cu3} = -3.5‰$ and $\varepsilon_{su} = 2.17‰$, and it can be checked that the balanced failure mode appears when the steel compression reinforcement are in plasticity, as:

$$0.330 = \frac{\delta'\varepsilon_{cu3}}{\varepsilon_{su}+\varepsilon_{cu3}} \leq \alpha_{bal} = \frac{-\varepsilon_{cu3}}{\varepsilon_{su}-\varepsilon_{cu3}} = 0.617 \qquad [1.45]$$

where δ' is chosen to be equal to a typical value of $\delta'=d'/d=0.125$. Figure 1.3 shows the sensitivity of the resistant moment with respect to the optimization parameter θ.

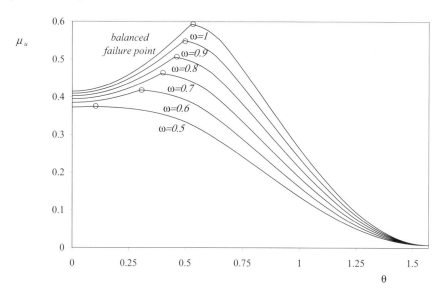

Figure 1.3. *Optimization of the reduced moment μ_u with respect to the steel reinforcement ratio $\omega_2/\omega_1 = \tan^2\theta$ for a given quantity $\omega \in \{0.5; 0.6; 0.7; 0.8; 0.9; 1\}$; $\delta' = 0.125$; C30-37 concrete and B500B steel*

In this case, the optimized solution corresponds to the balance failure point, as also remarked by Sieffert [SIE 10] (see also [THO 09] for instance). This optimized solution can be explicitly given by:

$$\theta_{bal} = \frac{1}{2}\arccos\left(\frac{\alpha_{bal}\psi}{\omega}\right) \quad \text{for} \quad \omega \geq \alpha_{bal}\psi \qquad [1.46]$$

Even if the design at the balance failure point is the optimum one from a strength point of view (at least for this case), for heavily reinforced sections, such a design is not necessarily optimal from the ductility point of view as we already discussed in this chapter (see also the discussion on this point in [PAR 75]). For lightly reinforced sections, the optimum design is obtained without compression steel reinforcement:

$$\theta = 0 \quad \text{for} \quad \omega \leq \alpha_{bal}\psi \qquad [1.47]$$

Figure 1.4 shows the evolution of the total steel content ω with respect to the solicitation μ_u.

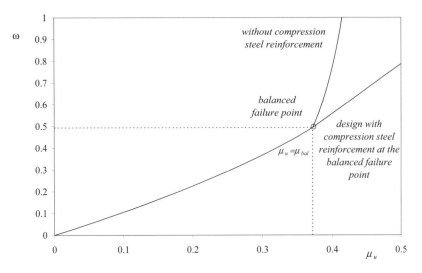

Figure 1.4. *Evolution of the normalized steel content ω with respect to the reduced moment μ_u for a reinforced concrete section with and without compression steel reinforcement;*

$$\omega = \frac{A_s f_{su}}{-f_{cu}bd} \; ; \; \delta' = 0.125$$

It is shown that adding some compression steel reinforcement designed at the balanced failure point may lead to a significantly more economical design. The optimization design can then be summarized as the following. For lightly reinforced section, it is not necessary to add compression steel reinforcement in the reinforced concrete section:

$$\omega = 1 - \sqrt{1 - 2\mu_u} \text{ and } \mu_u \leq \mu_{bal} = 0.8\alpha_{bal}(1 - 0.4\alpha_{bal}) \qquad [1.48]$$

For a heavily reinforced section, the optimized solution is obtained by adding some compression steel reinforcement in the reinforced concrete section, designed at the balance failure point as:

$$\omega(1 - \delta')\sin^2\left[\frac{1}{2}\arccos\left(\frac{\alpha_{bal}\psi}{\omega}\right)\right] = \mu_u - \mu_{bal} \text{ for}$$

$$\mu_u \geq \mu_{bal} = 0.8\alpha_{bal}(1 - 0.4\alpha_{bal}) \qquad [1.49]$$

The possible coincidence of the optimization solution with the balanced failure point depends on the type of reinforced concrete section considered, and especially on the values of α_{bal} with the geometrical parameter δ'. This sensitivity phenomenon is very similar to the optimization problem at SLS, where the optimization problem has been shown to be also very sensitive to the geometrical parameter δ', for some given concrete and steel characteristics. We will give below the critical boundary with respect to both α_{bal} and δ' that separates the two different behaviors for the optimization of the reinforcement bars.

When both the tensile and the compression reinforcements work in plasticity, it has been shown in equation [1.41] that the reduced moment in pivot B is represented by:

$$\mu_u = \omega\cos(2\theta)\left[1 - \frac{\omega\cos(2\theta)}{2}\right] + \omega(\sin^2\theta)(1-\delta') \text{ for } \frac{\delta'\varepsilon_{cu3}}{\varepsilon_{su} + \varepsilon_{cu3}}$$

$$\leq \alpha \leq \frac{-\varepsilon_{cu3}}{\varepsilon_{su} - \varepsilon_{cu3}} \qquad [1.50]$$

It is easy to show that this function has an optimum for:

$$\frac{\partial \mu_u}{\partial \theta} = 0 \Rightarrow \cos(2\theta) = \frac{1+\delta'}{2\omega} \Rightarrow \theta_{opt} = \frac{1}{2}\arccos\left(\frac{1+\delta'}{2\omega}\right) \qquad [1.51]$$

Therefore, two cases appear. In the first case, the optimization problem coincides with the value of the balanced failure point ($\alpha = \alpha_{bal}$ for $\delta' \geq \delta'_c$), because:

$$\theta_{bal} = \frac{1}{2}\arccos\left(\frac{\alpha_{bal}\psi}{\omega}\right) \geq \theta_{opt} = \frac{1}{2}\arccos\left(\frac{1+\delta'}{2\omega}\right)$$

$$\Rightarrow \delta' \geq \delta'_c = 2\psi\alpha_{bal} - 1 \qquad [1.52]$$

For instance, for a C30-37 type concrete and B500B steel, this condition is automatically verified as $\delta'_c = 2\psi\alpha_{bal} - 1 = -0.013 \leq 0$. This is exactly what appears in Figure 1.3, where it is shown numerically that the optimization problem of the strongest reinforced section exactly coincides with a design at the balanced failure point. Equation [1.52] analytically confirms this result.

However, in the case, for instance, of a *B400B* steel (typically for low steel characteristics), we find $\delta_c' = 2\psi\alpha_{bal} - 1 = 0.0689 \geq 0$, and this condition is verified only if $\delta' \geq \delta_c' = 0.0689$. In the second case, the optimization problem leads to a value that is different from the balanced failure point, as:

$$\theta_{bal} = \frac{1}{2}\arccos\left(\frac{\alpha_{bal}\psi}{\omega}\right) \leq \theta_{opt} = \frac{1}{2}\arccos\left(\frac{1+\delta'}{2\omega}\right)$$

$$\Rightarrow \delta' \leq \delta_c' = 2\psi\alpha_{bal} - 1 \qquad\qquad\qquad [1.53]$$

In this last case, the strongest reinforced section is not designed exactly at the balanced failure point ($\alpha = (1+\delta')/2\psi \neq \alpha_{bal}$ for $\delta' \leq \delta_c'$).

1.1.2. Simplified rectangular behavior – T-cross-section

1.1.2.1. Introduction – T-cross-section at ULS

The design at ULS of a reinforced concrete T-cross-section is now investigated based on the use of the rectangular simplified law for the compression concrete block, and a bilinear law for the steel that accounts for the hardening behavior. The reinforced concrete cross-section is shown in Figure 1.5.

A T-cross-section is analyzed where both compression (with area A_{s2}) and tensile (with area A_{s1}) steel bars reinforce the composite cross-section. The geometry of the cross-section is characterized by different length parameters b, b_w, h, h_0, d, d'. b is the width of the concrete slab, h_0 is the depth of the flange (slab) thickness. The width of the web is denoted by b_w. The position of the neutral axis is as usual, characterized by $y = \alpha d$ from the upper fiber of the cross-section.

For a general reinforced concrete section, the bending moment and normal force equilibrium equations are written with respect to the center of gravity of the tensile steel reinforcement, as:

$$\begin{cases} M_u = M_c - A_{s2}\sigma_{s2}(d-d') \\ N_u = N_c + \sigma_{s1}A_{s1} + \sigma_{s2}A_{s2} \end{cases} \text{with } y = \alpha d \qquad [1.54]$$

The calculation of the internal actions in the compression concrete block depends on the position of the neutral axis. If the depth of the compression block is within the flanged portion of the beam, that is the size of the compression block $0.8\alpha d$ is less than the flange (slab) thickness h_0, measured from the top of the slab ($0.8y = 0.8\alpha d < h_0$), then the section can be calculated as an "equivalent" rectangular cross-section with the width equal to b (as tensile concrete contribution is neglected in the analysis). Then, we find again the configuration previously investigated, for the design of reinforced concrete beams with rectangular cross-section at the ULS.

$$\frac{M_c}{-f_{cu}} = 0.8b\alpha d(d - 0.4\alpha d) \text{ and } \frac{N_c}{-f_{cu}} = 0.8b\alpha d \text{ for } 0.8\alpha d \leq h_0 \quad [1.55]$$

However, when the size of the compression block is larger than the flange (slab) thickness ($0.8y = 0.8\alpha d > h_0$), the calculation has to be based on the T-cross-section calculation.

$$\frac{M_c}{-f_{cu}} = (b - b_w)h_0\left(d - \frac{h_0}{2}\right) + 0.8b_w\alpha d(d - 0.4\alpha d) \text{ and}$$

$$\frac{N_c}{-f_{cu}} = (b - b_w)h_0 + 0.8b_w\alpha d \text{ for } 0.8\alpha d \geq h_0 \quad [1.56]$$

The transitory case between these two kinds of behavior is obtained for $y = \alpha d = h_0$ (see Figure 1.5).

$$\frac{M_T}{-f_{cu}} = bh_0\left(d - \frac{h_0}{2}\right) \text{ and } \frac{N_T}{-f_{cu}} = bh_0 \text{ for } 0.8\alpha d = h_0 \quad [1.57]$$

These characteristic values correspond to the total compression of the flanged portion of the cross-section, and strictly speaking, the neutral axis position is already in the web part of the T-cross-section as $y = \alpha d = h_0/0.8$.

The distinction between the two types of calculation (rectangular cross-section calculation or T-cross-section calculation) can be made from the following criterion:

$$M_c \leq M_T \quad \rightarrow \quad \text{Calculation based on a rectangular cross-section}$$

$$M_c \geq M_T \quad \rightarrow \quad \text{Calculation based on a T-cross-section} \quad [1.58]$$

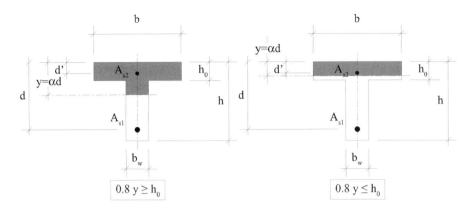

Figure 1.5. *Position of the neutral axis at ultimate limit state; T-cross-section – simplified rectangular constitutive law*

1.1.2.2. *Simplified rectangular behavior – T-cross-section with only tensile steel reinforcement*

The case of a reinforced T-cross-section with only tensile steel reinforcement is first presented. In this case, the bending moment equilibrium equation leads to $M_u = M_c$, and the condition for the type of calculation is reduced to:

$$M_u \leq M_T \quad \rightarrow \quad \text{Calculation based on a rectangular cross-section}$$

$$M_u \geq M_T \quad \rightarrow \quad \text{Calculation based on a T-cross-section} \qquad [1.59]$$

It is useful to note that these conditions are dependent on the solicitation M_u, which is different from the criterion used at the SLS for distinguishing a rectangular and T-cross-section calculation that was found to be independent of the solicitation.

The case $M_u \leq M_T$ associated with a rectangular cross-section with the width equal to b having been treated above, we will essentially focus the analysis on the other one based on $M_u \geq M_T$. Introducing $M_u = M_c$ in equation [1.56] leads to the modified reduced moment equation:

$$\mu_{u,T} = 0.8\alpha(1-0.4\alpha)\,\text{with}\,\mu_{u,T} = \frac{M_u - \left(1-\dfrac{b_w}{b}\right)M_T}{-b_w d^2 f_{cu}} \qquad [1.60]$$

where $\mu_{u,T}$ is the modified reduced moment. We recognize a second-order equation with respect to the position of the neutral axis α:

$$\alpha^2 - \frac{5}{2}\alpha + \frac{25}{8}\mu_{u,T} = 0 \qquad [1.61]$$

whose solution of interest is given by:

$$\alpha = 1.25(1-\sqrt{1-2\mu_{u,T}}) \qquad [1.62]$$

Once the position of the neutral axis is calculated, the tensile steel area is obtained from the normal force equilibrium equation:

$$A_{s1} = \frac{N_u - f_{cu}\left[(b-b_w)h_0 + 0.8b_w\alpha d\right]}{\sigma_{s1}} \qquad [1.63]$$

In the case of pivot A, for $\alpha \le \alpha_{AB}$ (where the neutral axis at pivot AB is given by equation [1.1]), the strain capacity of the tensile steel reinforcement ε_{s1} is equal to ε_{ud}, and the steel stress σ_{s1} is equal to $q\varepsilon_{ud}+q'$ (see equation [3.91] [CAS 12]), leading to:

$$\alpha \le \alpha_{AB} \Rightarrow A_{s1} = \frac{N_u - f_{cu}[(b-b_w)h_0 + 0.8b_w\alpha d]}{q\varepsilon_{ud}+q'} \qquad [1.64]$$

In the case of pivot B, for $\alpha \ge \alpha_{AB}$, the strain of the tensile steel reinforcement ε_{s1} depends on the position of the neutral axis. If the tensile steel reinforcement behaves in plasticity, the tensile steel area has to be calculated from:

$$A_{s1} = \frac{N_u - f_{cu}[(b-b_w)h_0 + 0.8b_w\alpha d]}{q\dfrac{\alpha-1}{\alpha}\varepsilon_{cu3}+q'} \quad \text{if } \alpha \in [\alpha_{AB};\alpha_{bal}] \qquad [1.65]$$

If the tensile steel reinforcement behaves in elasticity, the tensile steel area is calculated from:

$$A_{s1} = \frac{N_u - f_{cu}[(b-b_w)h_0 + 0.8b_w\alpha d]}{E_s \dfrac{\alpha-1}{\alpha}\varepsilon_{cu3}} \text{ if } \alpha \in \left[\alpha_{bal}; \frac{h}{d}\right] \tag{1.66}$$

1.1.2.3. Simplified rectangular behavior – T-cross-section with both tensile and compression steel reinforcement

1.1.2.3.1. Simplified rectangular behavior – pivot A

The case of a general T-cross-section with both tensile and compression steel reinforcements is now analyzed. The compression steel reinforcements are placed at a distance d' from the upper fiber of the cross-section, with the associated dimensionless parameter δ', defined by $\delta' = d'/d$. In pivot A, for $\alpha \leq \alpha_{AB}$, the bending moment equilibrium equation is written for a general T-cross-section, as:

$$\begin{aligned}M_u = &-f_{cu}\left[(b-b_w)h_0\left(d-\frac{h_0}{2}\right)+0.8b_w\alpha d(d-0.4\alpha d)\right]\\ &-A_{s2}d(1-\delta')\left[-m\varepsilon_{ud}\frac{\alpha-\delta'}{1-\alpha}+m'\right]\end{aligned} \tag{1.67}$$

The dimensionless equation is then obtained as:

$$\mu_{u,T} = 0.8\alpha(1-0.4\alpha)+\rho_{s2}'(1-\delta')\left[\frac{-m}{f_{cu}}\varepsilon_{ud}\frac{\alpha-\delta'}{1-\alpha}+\frac{m'}{f_{cu}}\right]\text{ with}$$

$$\mu_{u,T} = \frac{M_u-\left(1-\dfrac{b_w}{b}\right)M_T}{-b_wd^2f_{cu}} \tag{1.68}$$

where the normalized compression steel quantity is calculated as $\rho_{s2}' = A_{s2}/b_wd$. The problem is identical to the problem based on the rectangular cross-section with the modified reduced moment introduced in equation [1.60].

The determination of the dimensionless position of neutral axis α is then obtained from the resolution of a cubic equation:

$$\alpha^3 - \frac{7}{2}\alpha^2 + \alpha\left[\frac{25}{8}\mu_{u,T} + \frac{5}{2} + \frac{25}{8}\frac{\rho'_{s2}(1-\delta')}{(-f_{cu})}(m\varepsilon_{ud} + m')\right]$$
$$-\left[\frac{25}{8}\mu_{u,T} + \frac{25}{8}\frac{\rho'_{s2}(1-\delta')}{(-f_{cu})}(m\varepsilon_{ud}\delta' + m')\right] = 0 \qquad [1.69]$$

Equation [1.69] can be solved using Cardano's method (see Appendix 1 for the application of Cardano's method). The calculation of the parameters m and m' is the same as in the case of the rectangular cross-section. The tensile steel area is computed from the normal force equilibrium equation written as:

$$\alpha \leq \alpha_{AB} \Rightarrow A_{s1} = \frac{N_u - f_{cu}[(b-b_w)h_0 + 0.8b_w\alpha d] - A_{s2}\left(-m\varepsilon_{ud}\dfrac{\alpha-\delta'}{1-\alpha} + m'\right)}{q\varepsilon_{ud} + q'} \qquad [1.70]$$

1.1.2.3.2. Simplified rectangular behavior – pivot B

In pivot B, for $\alpha \geq \alpha_{AB}$, the bending moment equilibrium equation is written for a general T-cross-section, as:

$$\mu_{u,T} = 0.8\alpha(1-0.4\alpha) + \rho'_{s2}(1-\delta')\left[\frac{m}{f_{cu}}\varepsilon_{cu3}\frac{\alpha-\delta'}{\alpha} + \frac{m'}{f_{cu}}\right] \quad \text{with}$$

$$\mu_{u,T} = \frac{M_u - \left(1 - \dfrac{b_w}{b}\right)M_T}{-b_w d^2 f_{cu}} \qquad [1.71]$$

where the normalized compression steel quantity is calculated as $\rho'_{s2} = A_{s2}/b_w d$. The problem is identical to the problem based on the rectangular cross-section with the modified reduced moment introduced in equation [1.60].

The determination of the dimensionless position of neutral axis α is then obtained from the resolution of a cubic equation:

$$\alpha^3 - \frac{5}{2}\alpha^2 + \alpha\left[\frac{25}{8}\mu_{u,T} + \frac{25}{8}\frac{\rho'_{s2}(1-\delta')}{(-f_{cu})}(m\varepsilon_{cu3} + m')\right]$$
$$-\left[\frac{25}{8}\frac{\rho'_{s2}(1-\delta')}{(-f_{cu})}m\varepsilon_{cu3}\delta'\right] = 0 \qquad\qquad [1.72]$$

Equation [1.72] can be solved using Cardano's method (see Appendix 1 for the application of Cardano's method). The calculation of the parameters m and m' is the same as in the case of the rectangular cross-section. The tensile steel area is computed from the normal force equilibrium equation written in pivot B, with the tensile steel reinforcement in the plasticity range as:

$$A_{s1} = \frac{N_u - f_{cu}[(b-b_w)h_0 + 0.8b_w\alpha d] - A_{s2}\left(m\varepsilon_{cu3}\dfrac{\alpha-\delta'}{\alpha} + m'\right)}{q\dfrac{\alpha-1}{\alpha}\varepsilon_{cu3} + q'} \qquad [1.73]$$

if $\alpha \in [\alpha_{AB};\alpha_{bal}]$

If the tensile steel reinforcement behaves in elasticity, the tensile steel area is calculated from:

$$A_{s1} = \frac{N_u - f_{cu}[(b-b_w)h_0 + 0.8b_w\alpha d] - A_{s2}\left(m\varepsilon_{cu3}\dfrac{\alpha-\delta'}{\alpha} + m'\right)}{E_s\dfrac{\alpha-1}{\alpha}\varepsilon_{cu3}}$$

if $\alpha \in \left[\alpha_{bal};\dfrac{h}{d}\right]$ \qquad\qquad [1.74]

1.1.3. Comparison of design between serviceability limit state and ultimate limit state

1.1.3.1. Ultimate limit state versus serviceability limit state

The design of a reinforced concrete section at both the SLS and the ULS is compared in this section, for a rectangular reinforced concrete section. As a simplification, the concrete law at ULS is the simplified rectangular one.

The constitutive law of the steel tensile reinforcement is elastic and perfectly plastic. The section is reinforced only by some tensile steel reinforcement $(A_{s2} = 0)$.

As we will see, it is not obvious that the ULS necessarily leads to a safer design than the SLS. It depends, in fact, on the solicitation ratio between the ultimate limit moment and the serviceability limit moment.

$$\gamma = \frac{M_u}{M_{ser}} \geq 1 \qquad\qquad [1.75]$$

The specific parameter values (M_u, M_{ser}), which lead to an exactly coincident steel area design at the SLS and at the ULS, will be denoted by $(M_{l,u}, M_{l,ser})$. As a definition, we have for these specific parameters:

$$A_{s1,lser} = A_{s1,lu} \qquad\qquad [1.76]$$

The tensile steel area at SLS $A_{s1,lser}$ is calculated assuming a pivot B design at SLS:

$$A_{s1,lser} = \frac{\alpha_{l,ser}^2 bd}{2\alpha_e \left(1 - \alpha_{l,ser}\right)} \text{ and}$$

$$\alpha_{l,ser} = 1.5 \left(1 - \sqrt{1 - \frac{8}{3}\mu_{l,ser}}\right) \quad \text{in pivot } B \text{ at SLS} \qquad [1.77]$$

The tensile steel area at ULS $A_{s1,lu}$ is calculated assuming a pivot A or pivot B design at ULS, before the balanced failure point (the tensile steel reinforcement remain in their plastic part):

$$A_{s1,lu} = \frac{-0.8\alpha_{l,u} bd f_{cu}}{f_{su}} \text{ with } \alpha_{l,u} = 1.25\left(1 - \sqrt{1 - 2\mu_{l,u}}\right) \qquad [1.78]$$

The moments at ULS and SLS can be expressed with respect to their dimensionless counterparts as:

$$M_{l,u} = -\mu_{l,u}bd^2 f_{cu} = \mu_{l,u}bd^2 \frac{\alpha_{cc}f_{ck}}{\gamma_c} \text{ and}$$

$$M_{l,ser} = -\mu_{l,ser}bd^2 f_{cs} = \mu_{l,ser}bd^2 0.6 f_{ck} \qquad [1.79]$$

where α_{cc} is chosen here to be equal to $\alpha_{cc} = 0.85$ (even if $\alpha_{cc} = 1$ can be used in the French National Annexes of Eurocode 2 for building design, but other values can be found in some other annexes), and $\gamma_c = 1.5$. It can be shown with these numerical values that the loading factor γ can be calculated with respect to the reduced moments as:

$$\gamma = \frac{M_{l,u}}{M_{l,ser}} = \frac{17\mu_{l,u}}{18\mu_{l,ser}} \Rightarrow \mu_{l,ser} = \frac{17\mu_{l,u}}{18\gamma} \text{ and } \alpha_{l,ser} = 1.5\left(1 - \sqrt{1 - \frac{68\mu_{l,u}}{27\gamma}}\right) \quad [1.80]$$

By expressing $\alpha_{l,u}$ and $\alpha_{l,ser}$ as a function of the ultimate reduced moment $\mu_{l,u}$ and γ in the tensile steel area equality equation [1.76], we obtain a nonlinear equation written as:

$$\frac{1.5^2\left[1 - \sqrt{1 - \dfrac{68\ \mu_{1,u}}{27\ \gamma}}\right]^2}{2\left[-0.5 + 1.5\sqrt{1 - \dfrac{68\ \mu_{1,u}}{27\ \gamma}}\right]} + (1 - \sqrt{1 - 2\mu_{1,u}})\frac{\alpha_e f_{cu}}{f_{su}} = 0 \qquad [1.81]$$

After some lengthy manipulations, it can be shown that this nonlinear equation is in fact equivalent to the resolution of a cubic equation:

$$a_3\,\mu_{l,u}^3 + a_2\,\mu_{l,u}^2 + a_1\,\mu_{l,u} + a_0 = 0 \qquad \text{with:}$$

$a_3 = 289(-17 f_{su}^2 + 24 f_{cu}^2 \gamma \alpha_e^2)^2$

$a_2 = 204 f_{cu}\,\gamma \alpha_e\,(-1445 f_{su}^3 + 1156 f_{cu} f_{su}^2 \alpha_e + 2958 f_{cu} f_{su}^2 \gamma \alpha_e - 2040 f_{cu}^2 f_{su}\,\gamma \alpha_e^2 - 576 f_{cu}^3 \gamma^2\,\alpha_e^3)$

$a_1 = 36 f_{cu}\,\gamma^2\,\alpha_e\,(2601 f_{su}^3 - 2312 f_{cu} f_{su}^2 \alpha_e - 9180 f_{cu} f_{su}^2 \gamma \alpha_e + 7752 f_{cu}^2 f_{su}\,\gamma \alpha_e^2 + 576 f_{cu}^3 \gamma^2\,\alpha_e^3)$

$a_0 = 5832 f_{cu}^2 f_{su}\,\gamma^4\,\alpha_e^2(9 f_{su} - 8 f_{cu}\,\alpha_e)$ $\qquad [1.82]$

Cardano's method can be used to compute the limit moment $\mu_{l,u}$ as a function of the material characteristics:

$$\mu_{l,u} = \mu_{l,u}\left(\alpha_e, \gamma, \frac{-f_{cu}}{f_{su}}\right)$$

[1.83]

Numerical values of this limit reduced moment are available in standard textbooks (see for instance [PER 09, PER 10]).

We finally obtain the design rules:

$\mu_u \leq \mu_{l,u}$ → Design controlled by the Ultimate Limit State

$\mu_u \geq \mu_{l,u}$ → Design controlled by the Serviceability Limit State   [1.84]

1.1.3.2. *A numerical comparison of the ultimate limit state versus serviceability limit state*

We consider the design of a rectangular reinforced concrete section in simple bending, with only tensile steel reinforcement. The concrete law at ULS is the simplified rectangular law. The constitutive law of the steel tensile reinforcement is elastic and perfectly plastic. The geometrical and material characteristics of the reinforced concrete section are represented by:

$b = 0.25$ m $d = 0.48$ m
$f_{c28} = 25$ MPa $f_{yk} = 400$ MPa
$f_{cs} = -15$ MPa $f_{cu} = -0.85*25/1.5$ MPa
$f_{ss} = 400$ MPa $f_{su} = 400/1.15$ MPa
$E_s = 2.10^5$ MPa $\alpha_e = 15$
$M_u = 0.035$ MN.m $M_{ser} = 0.025$ MN.m   [1.85]

The loading parameter γ is calculated from $\gamma = M_u / M_{ser} = 1.4$. Using equation [1.82], the cubic equation for the determination of $\mu_{l,u}$ is written as:

$$8.410418114...\mu_{l,u}^3 + 1402.041361...\mu_{l,u}^2 - 987.997968...\mu_{l,u}$$
$$+ 169.9792469... = 0 \quad [1.86]$$

Cardano's method is used to compute the roots of the cubic (see Appendix 1). The cubic equation can be expressed in the canonical format $y^3 + py + q = 0$ leading to the canonical parameters $p = -9380.76232$ and $q = 349707.353$. It appears that the discriminant $p^3/27 + q^2/4 = -77163.0738$ is negative, which means that the three roots are real numbers. The solution of interest is finally equal to:

$$\mu_{l,u} = 0.299935 \tag{1.87}$$

The associated limit values are computed as $\mu_{lser} = 0.202337377$, $\alpha_{lu} = 0.459302931$, $\alpha_{lser} = 0.482171068$ and $\varepsilon_{s1l,u} = 4.12024311\text{‰}$. As $\varepsilon_{su} = f_{su}/E_s = 1.739130435\text{‰}$, then $\sigma_{s1l,u} = f_{su}$. Figure 1.6 shows the evolution of the limit reduced moment $\mu_{l,u}$ with respect to the loading parameter γ, crossing the point with the coordinates $(\mu_{l,u}, \gamma) = (0.299935; 1.4)$. Note that the complex nonlinear curve is, in fact, almost linear, and can be accurately linearized around the reference point, starting from:

$$f(\mu_{l,u}, \gamma) = a_3(\gamma)\mu_{l,u}^3 + a_2(\gamma)\mu_{l,u}^2 + a_1(\gamma)\mu_{l,u} + a_0(\gamma) = 0 \tag{1.88}$$

The linearization process is based on:

$$df = (\partial f / \partial \mu_{l,u})d\mu_{l,u} + (\partial f / \partial \gamma)d\gamma$$
$$\Rightarrow d\mu_{l,u}/d\gamma = -(\partial f / \partial \gamma)/(\partial f / \partial \mu_{l,u}) \tag{1.89}$$

We numerically obtain:

$$d\mu_{l,u}/d\gamma = 0.332640713 \Rightarrow$$
$$\mu_{l,u} - 0.299935407 = 0.332640713(\gamma - 1.4) \tag{1.90}$$

The linearization finally can be presented in the simple format:

$$\mu_{1,u} = 0.333\gamma - 0.166 \tag{1.91}$$

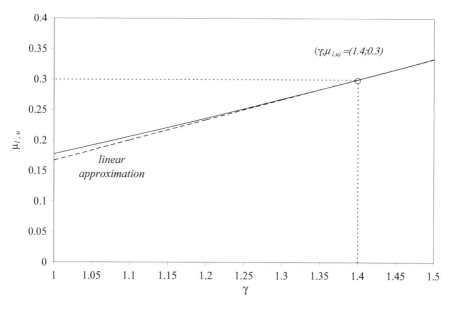

Figure 1.6. *Evolution of the limit moment* $\mu_{l,u}$ *versus the loading parameter* γ; *comparison between the exact solution and the linearized one around the point* $(\mu_{l,u}, \gamma) = (0.299935; 1.4)$

In the present case, with $\gamma = 1.4$ and $M_u = 0.035$ MN.m, $\mu_u = 0.042892157 < \mu_{l,u} = 0.299935$, which means that the ULS controls the design of this reinforced concrete section. The position of the neutral axis is calculated from $\alpha_u = 1.25(1 - \sqrt{1 - 2\mu_u}) = 0.054817165$, and $\sigma_{s1u} = 347.826087$ MPa. The tensile steel area is finally obtained for this reinforced concrete section from $A_{s1,u} = \dfrac{-0.8\alpha_u b d f_{cu}}{f_{su}} = 2.14335114$ cm². The final design can be based on $2\phi12$ ($A_{s1} = 2.262$ cm²). It is then easy to check the SLS, by computing the steel content $\rho_{s1} = 2.83 \times 10^{-2}$, the neutral axis position $\alpha_{ser} = -\rho_{s1} + \sqrt{\rho_{s1}(\rho_{s1} + 2)} = 0.211$ and finally the stresses in the tensile steel reinforcement $\sigma_{s1,ser} = 248$ MPa $< f_{ss} = 320$ MPa, and the one in the upper concrete fiber compression $\sigma_{c,sup,ser} = -4.42$ MPa $> f_{cs} = -15$ MPa.

1.1.4. *Biaxial bending of a rectangular cross-section*

1.1.4.1. *The design of a reinforced concrete section in biaxial bending*

This section is devoted to the design of a rectangular reinforced concrete section at ULS solicited by some normal forces and some biaxial bending solicitation. The concrete constitutive law at ULS is the simplified rectangular constitutive law. The section is only composed of tensile steel reinforcements (see Figure 1.7). We assume an elastic and perfectly plastic law for the steel reinforcements. The problem consists of designing this reinforced concrete section at the ULS, through a specific engineering example (Figure 1.7).

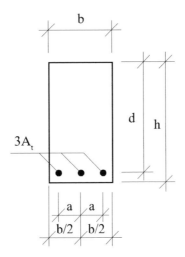

Figure 1.7. *Definition of the reinforced concrete cross-section*

The rectangular section is defined by both its width $b = 0.4$ m and its total depth $h = 0.7$ m. The section is reinforced by three equal tensile steel reinforcements for each area of A_t, which are unknown quantities to be designed. These tensile steel reinforcements are symmetrical with respect to the vertical axis of the rectangular cross-section. Their positions are defined by the length parameters $a = 0.16$ m and $d = 0.60$ m. The steel reinforcement has a characteristic yield strength $f_{yk} = 400$ MPa. The concrete has a uniaxial characteristic compressive strength $f_{ck} = 30$ MPa and the rectangular simplified law is used for the concrete behavior at ULS ($C30$-37-type).

The external screw action is equivalent to a single loading force $\overrightarrow{N} \to \mathrm{N}$ that acts at a point P, as shown in Figure 1.8. The section is, in fact, solicited by some normal force and biaxial bending as a result of the biaxial eccentricities of the applied axial force.

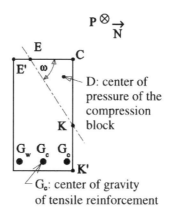

Figure 1.8. *Definition of the equivalent eccentric loading mode, with eccentricities in both directions*

As a consequence of the biaxial bending nature of the loading mode, the neutral axis is inclined with an angle ω with respect to the horizontal line as shown by Figure 1.8. It is assumed that the internal force $\overrightarrow{F_C}$ (equivalent axial force in the triangular compression concrete block), is located in a point *D*, which is the centre of pressure of the compression concrete part. The three steel tensile reinforcements are located in the points G_w (the subscript "w" refers to west), G_C (the subscript "c" refers to the centre) and G_e (the subscript "e" refers to the east), which are the centres of gravity of each steel reinforcement associated to the internal axial forces denoted by $\overrightarrow{F_{stw}}$, $\overrightarrow{F_{stc}}$ et $\overrightarrow{F_{ste}}$. It can be assumed that: $\overrightarrow{F_{st}} = \overrightarrow{F_{stw}} + \overrightarrow{F_{stc}} + \overrightarrow{F_{ste}}$. As shown by Figure 1.8, the points E and K are located on the segments CE' and CK' issued of the corner point C and bordering the rectangle.

According to Eurocode 2 rules, the ultimate limit strength is defined by:

$$f_{cu} = \frac{-\alpha_{cc} f_{ck}}{\gamma_c} \tag{1.92}$$

In Eurocode 2 rules, α_{cc} can be chosen to be equal to unity $(\alpha_{cc} = 1)$, at least in the French National Annexes, for building applications, in the case of rectangular or T-cross-sections in simple bending (constant or piecewise constant width section). It is required to reduce this factor α_{cc} to 10% if the width of the compression area decreases in the direction of the most compressed fiber, as it is the case in the considered example (with a triangular compression block). With respect to Eurocode 2 rules applied with the French National Annexes (see, for instance, [PER 09]), it would then be required to use $\alpha_{cc} = 0.9$ for this problem in the case of building applications. If now, we use $\alpha_{cc} = 0.85$ as required, for instance, for bridge designs in Eurocode 2, the reduced value can be approximately chosen equal to $\alpha_{cc} = 0.8$ to take into account the unfavorable triangular shape effect of the compression block. It is worth mentioning that this typical value of $\alpha_{cc} = 0.85$, which is reduced to $\alpha_{cc} = 0.8$ in the area of the compression block where the width is decreasing in the direction of the most compressed fiber, is the value recommended in the old French rules (*Béton Armé aux Etats Limites*). In the present case, We calculate for the C30-37-type concrete $f_{cu} = -0.8 \times 30 / 1.5 = -16$ MPa.

As shown in Figure 1.9, the origin of the frame is chosen at the point G_C.

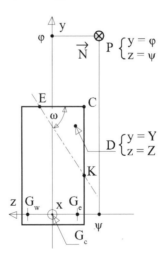

Figure 1.9. *Geometrical parametrization of the rectangular reinforced cross-section, with respect to both the tensile reinforcements and the loading mode*

It is possible to define the coordinates of each characteristic points in Figure 1.9 in the frame (G_c, x, y, z). The axial force $\vec{N}$, is applied at the point P whose coordinates are $(0, \varphi, \psi)$. The centre of pressure of the concrete triangular compression block is denoted by the point D defined by the coordinates $(0, Y, Z)$. The points G_W, G_C and G_e have the following coordinates: $(0,0,a)$, $(0,0,0)$ and $(0,0,-a)$. The tensile steel reinforcement in G_W is the most solicited in such a loading configuration. A more comprehensive representation is shown in Figure 5.10 with a three-dimensional perspective.

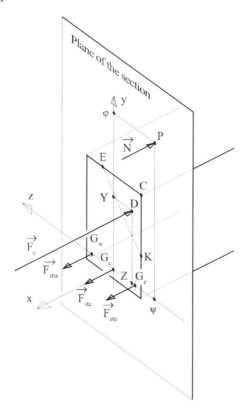

Figure 1.10. *Three-dimensional representation of the reinforced concrete cross-section*

For an easier resolution of the problem, the design is decomposed into different steps before the final steel area calculation.

1) The effective depth is the distance between the most compressed fiber in the concrete and the center of gravity of the most tensioned steel reinforcement. Calculate the effective depth d'' with respect to a, b, d and ω.

2) It is possible to define the relative depth α, which is a positive dimensionless parameter, as the ratio between the distance from the neutral axis to the most compressed concrete fiber y_u and the effective depth d''. Prove that $F_C = \dfrac{(0.8.\alpha\, d'')^2}{\sin 2\omega} f_{cu}$. Determine the coordinates $(0,Y,Z)$ of the point D, center of pressure of the triangular compression block, with respect to b, d, d'', α and ω. Show that $F_C = 4.5\,(d - Y)\left(\dfrac{b}{2} + Z\right) f_{cu}$.

3) In the following, it is assumed that the strains in each tensile steel reinforcement is larger than $\varepsilon_{su} = \dfrac{f_{su}}{E_s}$ (all the tensile steel reinforcements behave in the plasticity range). Show why the screw internal forces in the tensile steel reinforcements is equivalent to single force $\vec{F_{st}}$ applied in G_C. Write the equilibrium equations (axial force equilibrium equation in the x direction, moment equilibrium equation with respect to y and moment equilibrium equation with respect to z) using the variables F_C, F_{st}, N, Y, Z, φ and ψ."

4) Prove that F_C can be obtained by solving a cubic equation expressed with a t variable: $t^3 - 2.25\,f_{cu}bdt^2 + 2.25\,f_{cu}(\varphi b - 2\,\psi d)Nt + 4.5\,f_{cu}\varphi\psi N^2 = 0$. Determine, using Cardano's method, the numerical values of the roots of this cubic and select the one that is of physical interest in our design problem. In the following, we will consider the numerical values $N = -0.12\ MN$, $\varphi = 1.05$ m and $\psi = -0.3$ m.

5) Explicit Y and Z with respect to F_C, N, φ and ψ, and then express $\tan \omega$ with respect to b, d, Y and Z, in order to obtain α with respect to F_C, ω, d'' and f_{cu}.

6) Express analytically A_t with respect to N, F_C and f_{su}. Calculate the steel tensile reinforcement area for the design of this reinforced concrete section at its ULS.

1.1.4.2. *Resolution*

1) According to Figure 1.11, we calculate:

$$d'' = (a+b/2) \times \sin\omega + d \times \cos\omega \qquad [1.93]$$

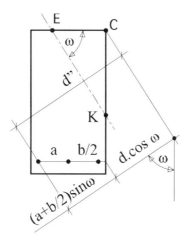

Figure 1.11. *Biaxial bending of a reinforced concrete section; geometrical analysis*

2) Using Figure 1.12, we calculate:

$$CH' = 0.8 \times CH = 0.8 \times \alpha \times d'' \text{ and}$$

$$CE' = \frac{CH'}{\sin\omega}; \quad CK' = \frac{CH'}{\cos\omega} \qquad [1.94]$$

In the rectangle triangle $CE'K'$, the stress is constant and is equal to $f_{cu.}$ The force F_C is calculated from the area of this triangle:

$$F_C = \frac{CH'^2 \times f_{cu}}{2 \times \sin\omega \times \cos\omega} = \frac{(0.8 \times \alpha \times d'')^2 \times f_{cu}}{\sin(2\,\omega)} \qquad [1.95]$$

It can be remarked that the shape filling coefficient (denoted by ψ in the other chapters) is equal for this biaxial loading problem to $(0.8)^2$. As the stress distribution is uniform in the triangle $CE'K'$ with the simplified rectangle constitutive law, the center of pressure in the compression block coincides with the center of gravity of the triangular compression block, characterized by:

$$\vec{G_cD} = \vec{G_cC} + \frac{\vec{CE'} + \vec{CK'}}{3} = (0, d, \frac{-b}{2})$$

$$+ (0, 0, \frac{0.8 \times \alpha \times d''}{3sin\ \omega}) + (0, \frac{-0.8 \times \alpha \times d''}{3cos\ \omega}, 0) =$$

$$(0, d - \frac{0.8 \times \alpha \times d''}{3 \times cos\ \omega}, \frac{-b}{2} + \frac{0.8 \times \alpha \times d''}{3 \times sin\ \omega}) = (0, Y, Z) \qquad [1.96]$$

We finally obtain the normal force in the compression block as:

$$\frac{0.8 \times \alpha \times d''}{cos\ \omega} = 3(d - Y); \frac{0.8 \times \alpha \times d''}{sin\ \omega} = 3(\frac{b}{2} + Z) \text{ and}$$

$$F_C = 4.5\ (d - Y)(0.5b + Z) f_{cu}. \qquad [1.97]$$

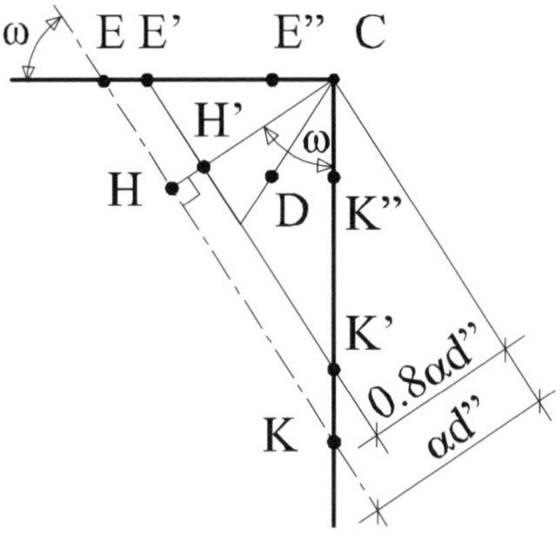

Figure 1.12. *Geometrical analysis inside the triangular compression block*

3) As the tensile steel reinforcements are in the plasticity stage, we have $\varepsilon_{sti} > \varepsilon_{su} = \frac{f_{su}}{E_s}$, and then the stress in the steel reinforcement is the one of the plateau $\sigma_{sti} = f_{su}\ \forall i \in \{w,c,e\}$. The area of the steel reinforcements is assumed to be equal and then the internal forces field is symmetrical with respect to the plane (xG_cy), even if the strain values in each reinforcements potentially differ. The internal screw is equivalent to a single axial load acting at the centre of gravity of the tensile steel reinforcement G_C. The three

equilibrium equations (one axial force equilibrium equation and two moment equilibrium equations) are then written as:

$$N = F_C + F_{st}; \ N\psi = F_C Z \ \text{and} - N\varphi = -F_C Y \qquad [1.98]$$

4) Using equation [1.93] and the equilibrium equation [1.94], the normal force calculation in concrete can be expressed by:

$$F_C = 4.5 \ (d - \frac{N\varphi}{F_C}) \ (\frac{b}{2} + \frac{N\psi}{F_C}) f_{cu} \qquad [1.99]$$

We finally obtain a cubic equation that is independent of the pivot considered:

$$F_C^3 - \frac{4.5}{2} bdf_{cu} F_C^2 + \frac{4.5}{2} (\varphi b - 2 \psi d) \ N f_{cu} F_C + 4.5 \varphi \psi f_{cu} N^2 = 0 \qquad [1.100]$$

This cubic equation $t^3 + 8.64t^2 + 3.3696t + 0.326592 = 0$ expressed in terms of the variable t ($t = F_c$) can be solved using Cardano's method (see Appendix 1), with the reduced parameters calculated as $p = -21.5136$ MN^2 and $q = 38.3979$ MN^3 leading to the discriminant $4p^3 + 27q^2 = -220.2463$ MN^6. The three real roots of the cubic equation are $F_{c1} = -8.2357$ MN, $F_{c2} = -0.2370$ MN and $F_{c3} = -0.1673$ MN.

5) Y, Z, $\tan\omega$ and α are calculated as:

$$Y = \frac{N\varphi}{F_C}; \ Z = \frac{N\psi}{F_C}; \ \tan \omega = \frac{\sin \omega}{\cos \omega} = \frac{d - Y}{0.5b + Z};$$

$$d^{\prime} = (a + b/2) \times \sin \omega + d \times \cos \omega,$$

$$\alpha = \frac{1}{0.8d^{\prime\prime}} \sqrt{\frac{\sin 2\omega F_C}{f_{cu}}} \qquad [1.101]$$

6) The numerical results are presented in Table 1.1. The calculation of both the normal force in concrete F_C and the dimensionless position of the neutral axis α is independent of the pivot considered. However, the strain values in each steel reinforcement depends on the pivot considered, namely pivot A or pivot B. For pivot A behavior, the strains are calculated as:

$$\varepsilon_{stw} = \varepsilon_{ud}; \; \varepsilon_{stc} = \varepsilon_{ud} \left[1 - \frac{a\sin\omega}{(1-\alpha)d''} \right] \text{ and}$$

$$\varepsilon_{ste} = \varepsilon_{ud} \left[1 - \frac{2a\sin\omega}{(1-\alpha)d''} \right] \quad \text{in pivot } A \qquad [1.102]$$

For pivot B, the strains are calculated from:

$$\varepsilon_{stw} = \frac{-\varepsilon_{cu3}}{\alpha}(1-\alpha), \; \varepsilon_{stc} = \frac{-\varepsilon_{cu3}}{\alpha}\left(1-\alpha-\frac{a\sin\omega}{d''}\right), \; \varepsilon_{ste} = \frac{-\varepsilon_{cu3}}{\alpha}\left(1-\alpha-\frac{2a\sin\omega}{d''}\right) \quad [1.103]$$

When using Eurocode 2 with elastoplastic steel reinforcements (neglecting the hardening phenomenon), it is possible to assume very large ductility, which means that pivot A, in fact, disappears with this constitutive law at ULS. Therefore, the strains will be computed only with a pivot B rule in Table 1.1. Furthermore, we have:

$$CE = \frac{\alpha d''}{\sin\omega} \text{ and } CK = \frac{\alpha d''}{\cos\omega} \qquad [1.104]$$

i	1	2	3
F_c (MN)	−8.23566798	−0.23702707	−0.16730493
Y (m)	0.015299305	0.531584843	0.753115852
Z (m)	−0.00437123	−0.151881384	−0.215175958
$\tan\omega$	2.988827742	1.421802254	10.08937
ω (rad)	1.24792479	0.95783716	1.472004765
$\omega°$	71.50082361	54.88002671	84.33966044
d'' (m)	0.531772773	0.639635974	0.417423235
α	1.308267544	0.230747955	0.135671301
Pivot	Pivot B	Pivot B	Pivot B
CE (m)	0.733607887	0.180444811	0.056909841
CK (m)	2.192627605	0.256556839	0.574184444
ε_{ste} (‰)	−2.35140738	5.461181156	2.617392814
ε_{stc} (‰)	−1.588056809	8.564623121	12.45751869
ε_{stw} (‰)	−0.824706238	11.66806509	22.29764456
σ_{ste} (MPa)	−347.826 087	347.826 087	347.826 087
σ_{stc} (MPa)	−317.6113617	347.826 087	347.826 087
σ_{stw} (MPa)	−164.9412475	347.826 087	347.826 087

Table 1.1. *Numerical values for the design of reinforced concrete section with biaxial bending*

The three solutions of the cubic equation [1.100] are referred to solution 1, solution 2 and solution 3. Solution 1 is not admissible as some steel reinforcement bars are in the elasticity regime, which is not compatible with the fundamental assumptions of the plasticity regime for each bar. Solution 3 is not admissible because the center of pressure is out of the section in this case $Y \geq h$, which is not admissible. Solution 2 is the only admissible solution for this problem. The normal force equilibrium equation is written as:

$$N = 3A_{st,cal} f_{su} + F_c \qquad [1.105]$$

The steel area of each bar is then obtained from $A_{st,cal} = 1.1215 \ cm^2$. This section can be reinforced by adding $3\phi12 = (3.393 \ cm^2)$, to support the biaxial bending solicitation in addition to the axial compression.

1.2. ULS – extended analysis

1.2.1. *Bilinear constitutive law for concrete – rectangular cross-section*

1.2.1.1. *Data of the problem*

This section is devoted to the design of a rectangular reinforced concrete section at ULS solicited in simple bending. The concrete constitutive law at ULS is the bilinear constitutive law (similar to an elastic and perfectly plastic constitutive law). The section is composed of tensile and compression steel reinforcements (see Figure 1.1). We assume a plasticity hardening law for the steel reinforcements. The problem consists of designing the tensile steel reinforcement of the reinforced concrete section at the ULS, for a given bending moment solicitation. The geometrical parameters of the problem are given by:

$$b = 0.2 \ \text{m}; \ d = 0.40 \ \text{m}; \ d' = 0.05 \ \text{m}; \ h = 0.47 \ \text{m};$$

$$A_{s2} = 0.848 \ \text{cm}^2; \ A_{s1}: \text{unknown} \qquad [1.106]$$

The compression and tensile steel reinforcements are made up of steel B500B steel bars. Hardening effect is taken into account. The steel parameters at ULS are:

$$E_s = 200 \ \text{GPa}; \varepsilon_{ud} = 45\text{\textperthousand}; \varepsilon_{uk} = 50\text{\textperthousand}; f_{su} = 434.783 \ \text{MPa}; k = 1.08 \quad [1.107]$$

The ductility is characterized by the coefficient k, and the ultimate limit strain ε_{uk}. The admissible limit strain for steel is equal to ε_{ud}. The plasticity parameters of this steel are calculated as:

$$q = \frac{(k-1)f_{su}}{\varepsilon_{uk} - \varepsilon_{su}} = 727.273 \text{ MPa}, q' = f_{su}\left(1 - \frac{q}{E_s}\right)$$

$$= 433.202 \text{ MPa and } \varepsilon_{su} = \frac{f_{su}}{E_s} = 2.17\text{‰} \qquad [1.108]$$

The concrete has a class C30/37-type and is modeled with a bilinear constitutive law, characterized by the characteristics values:

$$\varepsilon_{c3} = -1,75\text{‰} \; ; \; \varepsilon_{cu3} = -3,5\text{‰}; \; f_{cu} = -20 \text{ MPa} \qquad [1.109]$$

The exercise consists of designing the tensile steel reinforcement area of this reinforced concrete section at the ULS for a given solicitation $M_u = 0.14$ MN.m. This design needs to determine in an analytical and numerical way the sectional resistant moment at the ULS calculated at the center of gravity of the tensile steel reinforcement $M_{res,u}(\alpha)$. This resistant moment is a function of the dimensionless relative depth α. This function is continuous and piecewise differentiable. For the parameter of interest, the resolution of the nonlinear function $M_{res,u}(\alpha) = M_u$ gives the parameter α associated with the working behavior of the cross-section, and then the tensile steel area can be calculated from the application of the normal force equilibrium equation.

1.2.1.2. Resolution – pivot A1

Pivot $A1$ is characterized by the domain of variation in the position of the neutral axis position:

$$\alpha \in [0; \alpha_{A_1 A_2}] \text{ with } \alpha_{A_1 A_2} = \frac{-\varepsilon_{c3}}{\varepsilon_{ud} - \varepsilon_{c3}} = \frac{1.75}{46.75} = 0.037433 \qquad [1.110]$$

The section behaves in pivot A, which means that the tensile steel reinforcements have reached their strain limit capacity $\varepsilon_{s1} = \varepsilon_{ud}$, and the concrete behaves in the elastic regime $\varepsilon_{c,sup} \in [\varepsilon_{c3}; 0]$. It can be shown for this range of parameters that the upper steel reinforcements (usually in compression but here in tension) are in the plasticity regime, as the strain in the upper steel reinforcement, at each bound of the interval $\alpha \in [0; \alpha_{A_1 A_2}]$, is in the plasticity branch:

$$\varepsilon_{s2}(\alpha = 0) = \delta' \varepsilon_{ud} = 5.625\text{‰} > \varepsilon_{su} = 2.17\text{‰} \quad \text{and}$$

$$\varepsilon_{s2}(\alpha = \alpha_{A_1 A_2}) = \frac{\alpha_{A_1 A_2} - \delta'}{\alpha_{A_1 A_2}} \varepsilon_{c3} = 4.09375\text{‰} > \varepsilon_{su} = 2.17\text{‰} \qquad [1.111]$$

In pivot A_1, the reinforcement in the upper part of the cross-section works in tension in the plasticity regime:

$$\sigma_{s2} = q\varepsilon_{s2} + q' \geq 0 \qquad\qquad\qquad [1.112]$$

Using equation [3.120] [CAS 12], the resistant moment calculated at the center of gravity of the tensile steel reinforcement is given for $\alpha \in [0; \alpha_{A_1 A_2}]$ as:

$$M_{res,u}(\alpha) = bd^2 f_{cu} \frac{1}{2} \frac{\alpha^2}{1-\alpha} \left(1 - \frac{\alpha}{3}\right) \frac{\varepsilon_{ud}}{\varepsilon_{c3}}$$

$$- A_{s2}(d - d') \left[q \frac{\alpha d - d'}{(\alpha - 1)d} \varepsilon_{ud} + q' \right] \qquad [1.113]$$

1.2.1.3. Resolution – pivot A2

Pivot $A2$ is characterized by the domain of variation for the position of the neutral axis:

$$\alpha \in [\alpha_{A_1 A_2}; \alpha_{AB}] \quad \text{with} \quad \alpha_{A_1 A_2} = \frac{-\varepsilon_{c3}}{\varepsilon_{ud} - \varepsilon_{c3}} = \frac{1.75}{46.75} = 0.037433 \quad \text{and}$$

$$\alpha_{AB} = \frac{-\varepsilon_{cu3}}{\varepsilon_{ud} - \varepsilon_{cu3}} = \frac{3.5}{48.5} = 0.072165 \qquad [1.114]$$

The section behaves in pivot A, which means that the tensile steel reinforcements have reached their strain limit capacity $\varepsilon_{s1} = \varepsilon_{ud}$, and the concrete behaves in the elastoplastic regime $\varepsilon_{c,\text{sup}} \in [\varepsilon_{cu3}; \varepsilon_{c3}]$. It can be shown for this range of parameters that the upper steel reinforcements are in the plasticity regime, as the strain in the upper steel reinforcement, at each bound of the interval $\alpha \in [\alpha_{A_1A_2} ; \alpha_{AB}]$ is in the plasticity branch:

$$\varepsilon_{s2}(\alpha = \alpha_{A_1A_2}) = \frac{\alpha_{A_1A_2} - \delta'}{\alpha_{A_1A_2}}\varepsilon_{c3} = 4.09375\text{‰} > \varepsilon_{su} = 2.17\text{‰}$$

$$\text{and } \varepsilon_{s2}(\alpha = \alpha_{AB}) = \frac{\alpha_{AB} - \delta'}{\alpha_{AB}}\varepsilon_{cu3} = 2.5625\text{‰} > \varepsilon_{su} = 2.17\text{‰} \qquad [1.115]$$

Equation [1.112] is also valid for the calculation of the stress in the upper steel reinforcement (usually in compression for positive bending moment).

More generally, the upper steel reinforcements are always in the tensile regime in pivot A for:

$$\delta' = \frac{d'}{d} \geq \alpha_{AB} \qquad [1.116]$$

It can be checked that this last inequality is valid in the reinforced concrete section studied here, as $\delta' = 0.125 \geq \alpha_{AB} = 0.0722$. Using equation [3.125] of [CAS 12], the resistant moment calculated at the centre of gravity of the tensile steel reinforcement is given for $\alpha \in [\alpha_{A_1A_2} ; \alpha_{AB}]$ as:

$$M_{res,u}(\alpha) = -bd^2 f_{cu}\left[\alpha - (1-\alpha)\left(\frac{-\varepsilon_{c3}}{\varepsilon_{ud}}\right)\right] \times \left[1 - \frac{\alpha}{2} + \frac{1-\alpha}{2}\left(\frac{-\varepsilon_{c3}}{\varepsilon_{ud}}\right)\right]$$

$$-bd^2 f_{cu}\frac{1-\alpha}{2}\left(\frac{-\varepsilon_{c3}}{\varepsilon_{ud}}\right)\left[\frac{2}{3}(1-\alpha)\left(\frac{-\varepsilon_{c3}}{\varepsilon_{ud}}\right) + 1 - \alpha\right]$$

$$-A_{s2}(d-d')\left[q\frac{\alpha d - d'}{(\alpha-1)d}\varepsilon_{ud} + q'\right] \qquad [1.117]$$

1.2.1.4. Resolution – pivot B

Pivot B is characterized by the domain of variation for the position of the neutral axis:

$$\alpha \in [\alpha_{AB} ; 1] \text{ with } \alpha_{AB} = \frac{-\varepsilon_{cu3}}{\varepsilon_{ud} - \varepsilon_{cu3}} = \frac{3.5}{48.5} = 0.072165 \qquad [1.118]$$

The compression concrete block has reached its ultimate strain capacity characterized by $\varepsilon_{c,sup} = \varepsilon_{cu3}$. Using equation [3.125] of [CAS 12], the resistant moment calculated at the center of gravity of the tensile steel reinforcement is given for $\alpha \in [\alpha_{AB};1]$ as:

$$M_{res,u}(\alpha) = -bd^2 f_{cu} \left[\alpha \left(1 - \frac{\alpha}{2}\right)\left(1 - \frac{\varepsilon_{c3}}{\varepsilon_{cu3}}\right) + \frac{\alpha}{2}\frac{\varepsilon_{c3}}{\varepsilon_{cu3}}\left(1 - \frac{\alpha}{3}\frac{\varepsilon_{c3}}{\varepsilon_{cu3}}\right) \right]$$

$$- A_{s2}(d - d')\left[m \frac{\alpha d - d'}{\alpha d}\varepsilon_{cu3} + m' \right] \qquad [1.119]$$

The value of the parameters m and m' depend on the dimensionless position of neutral axis α as:

$$m = q \text{ and } m = q \text{ for } \alpha \in \left[\alpha_{AB}; \frac{\delta' \varepsilon_{cu3}}{\varepsilon_{cu3} - \varepsilon_{su}} \right] = [0.072165; 0.07711]$$

$$m = E_s \text{ and } m' = 0 \text{ for } \alpha \in \left[\frac{\delta' \varepsilon_{cu3}}{\varepsilon_{cu3} - \varepsilon_{su}}; \frac{\delta' \varepsilon_{cu3}}{\varepsilon_{cu3} + \varepsilon_{su}} \right]$$

$$= [0.07711; 0.32992]$$

$$m = q \text{ and } m' = -q' \text{ for } \alpha \in \left[\frac{\delta' \varepsilon_{cu3}}{\varepsilon_{cu3} + \varepsilon_{su}}; 1 \right] \qquad [1.120]$$

$$= [0.32992; 1]$$

1.2.1.5. Synthesis – nonlinear problem to be solved

Whatever the pivot considered, it is observed that the nonlinear function $M_{res,u}(\alpha) = M_u$ is a cubic equation in general, in the presence of compression steel reinforcement.

We calculate the moment at the transition between pivot A and pivot B from equation [1.119], detailed below with $m = q$ and $m' = q'$:

$$M_{AB} = M_{res,u}\left(\alpha = \alpha_{AB}\right) = -bd^2 f_{cu}$$

$$\left[\alpha_{AB}\left(1-\frac{\alpha_{AB}}{2}\right)\left(1-\frac{\varepsilon_{c3}}{\varepsilon_{cu3}}\right)+\frac{\alpha_{AB}}{2}\frac{\varepsilon_{c3}}{\varepsilon_{cu3}}\left(1-\frac{\alpha_{AB}}{3}\frac{\varepsilon_{c3}}{\varepsilon_{cu3}}\right)\right]$$

$$-A_{s2}\left(d-d'\right)\left[q\frac{\alpha_{AB}d-d'}{\alpha_{AB}d}\varepsilon_{cu3}+q'\right] = 0.020754 \ MN.m \qquad [1.121]$$

It can be seen that $M_u = 0.14$ MN.m $> M_{AB} = 0.020754$ MN.m and the section behaves in pivot B. Moreover, it can be shown, according to Figure 1.13, that:

$$M_u = 0.14 \ \text{MN.m} \in \left[M_{res,u}\left(\alpha = \frac{\delta'\varepsilon_{cu3}}{\varepsilon_{cu3}-\varepsilon_{su}}\right); M_{res,u}\left(\alpha = \frac{\delta'\varepsilon_{cu3}}{\varepsilon_{cu3}+\varepsilon_{su}}\right)\right] \quad [1.122]$$

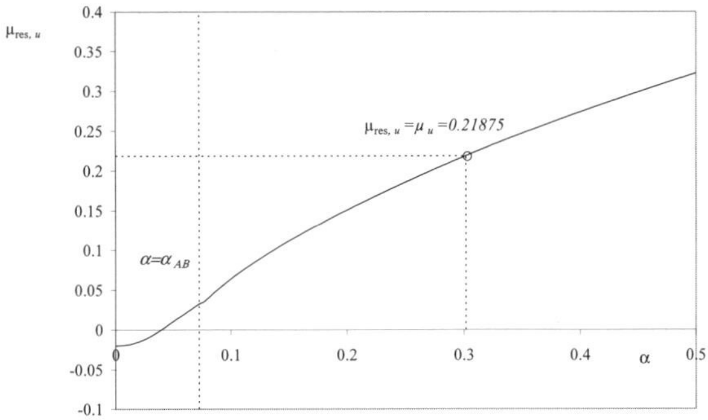

Figure 1.13. *Evolution of the reduced moment $\mu_{res,u}$ with respect to α; graphical resolution for $\mu_{res,u}(\alpha) = \mu_u = 0.21875; \ \alpha = 0.301717$*

The section behaves in pivot B with compression steel reinforcement in elasticity for $\mu_{res,u}(\alpha) = \mu_u = 0.21875$. The dimensionless position of the neutral axis has then to be calculated from the nonlinear equation [1.119] specialized with the compression steel reinforcement in the elasticity range, and written as:

$$M_u = -bd^2 f_{cu}\left[\alpha\left(1-\frac{\alpha}{2}\right)\left(1-\frac{\varepsilon_{c3}}{\varepsilon_{cu3}}\right)+\frac{\alpha}{2}\frac{\varepsilon_{c3}}{\varepsilon_{cu3}}\left(1-\frac{\alpha}{3}\frac{\varepsilon_{c3}}{\varepsilon_{cu3}}\right)\right]$$

$$-A_{s2}(d-d')E_s\frac{\alpha-\delta'}{\alpha}\varepsilon_{cu3} \qquad [1.123]$$

which can be written as a cubic equation:

$$-\frac{\alpha^3}{2}\left[1-\frac{\varepsilon_{c3}}{\varepsilon_{cu3}}+\frac{1}{3}\left(\frac{\varepsilon_{c3}}{\varepsilon_{cu3}}\right)^2\right]+\alpha^2\left(1-\frac{\varepsilon_{c3}}{2\varepsilon_{cu3}}\right)$$

$$+\alpha\frac{A_{s2}E_s d(1-\delta')\varepsilon_{cu3}+M_u}{bd^2 f_{cu}}-\frac{A_{s2}E_s d'(1-\delta')\varepsilon_{cu3}}{bd^2 f_{cu}}=0 \qquad [1.124]$$

The cubic equation is numerically described as:

$$a\alpha^3+b\alpha^2+c\alpha+d=0 \text{ with } a=-0.186667, b=0.48,$$
$$c=-0.119224 \text{ and } d=-0.002597 \qquad [1.125]$$

Cardano's method can be used as detailed in Appendix 1. The canonical parameters p and q are calculated as:

$$p=\frac{3ac-b^2}{3a^2}=-1.565382 \text{ and } q=\frac{27a^2 d+2b^3-9abc}{27a^3} \qquad [1.126]$$
$$=-0.698106$$

leading to $4p^3+27q^2=-2.18488<0$. The cubic has three real roots calculated as:

$$\begin{cases} \alpha_1=2\sqrt{\frac{-p}{3}}\cos\left[\dfrac{Arc\cos\left(\dfrac{3q}{2p}\sqrt{\dfrac{3}{-p}}\right)+2\pi}{3}\right]-\dfrac{b}{3a}=-0.020137 \\[4em] \alpha_2=2\sqrt{\frac{-p}{3}}\cos\left[\dfrac{Arc\cos\left(\dfrac{3q}{2p}\sqrt{\dfrac{3}{-p}}\right)+4\pi}{3}\right]-\dfrac{b}{3a}=0.301717 \\[4em] \alpha_3=2\sqrt{\frac{-p}{3}}\cos\left[\dfrac{Arc\cos\left(\dfrac{3q}{2p}\sqrt{\dfrac{3}{-p}}\right)}{3}\right]-\dfrac{b}{3a}=2.289848 \end{cases} \qquad [1.127]$$

The solution of the problem is $\alpha=\alpha_2=0.301717$, the other two solutions being negative or greater than unity. For this value of the neutral position

$\alpha = \alpha_2 = 0.301717$, the strain in the tensile and compression steel reinforcements can be calculated as:

$$\varepsilon_{s1} = \frac{\alpha-1}{\alpha}\varepsilon_{cu3} = 8.1‰ >> \varepsilon_{su} = 2.17‰ \text{ and } \varepsilon_{s2} = \frac{\alpha-\delta'}{\alpha}\varepsilon_{cu3}$$

$$= -2.05‰ > -\varepsilon_{su} = -2.17‰ \qquad [1.128]$$

It is easily checked that the tensile steel reinforcements work in plasticity whereas the compression steel reinforcements are in the elastic range. The steel reinforcement area is then computed from the normal force equilibrium equation (see also equation [3.133] of [CAS 12] for the component of the compression concrete block):

$$A_{s1}\sigma_{s1} + A_{s2}\sigma_{s2} + bdf_{cu}\alpha\left(1 - \frac{\varepsilon_{c3}}{2\varepsilon_{cu3}}\right) = 0 \qquad [1.129]$$

The tensile steel reinforcement area is then expressed from:

$$A_{s1} = -\frac{A_{s2}E_s}{\sigma_{s1}}\frac{\alpha-\delta'}{\alpha}\varepsilon_{cu3} - \frac{bdf_{cu}}{\sigma_{s1}}\alpha\left(1 - \frac{\varepsilon_{c3}}{2\varepsilon_{cu3}}\right)\text{with}$$

$$\sigma_{s1} = q\varepsilon_{s1} + q' = 439.093 \text{ MPa} \qquad [1.130]$$

We finally obtain $A_{s1} = 9.037 \text{ cm}^2$. The section can be designed with $3\phi20$ for the tensile steel reinforcement ($A_{s1} = 9.425 \text{ cm}^2$).

1.2.2. Parabola–rectangle constitutive law for concrete – rectangular cross-section

1.2.2.1. Data of the problem

In this section, the design of a rectangular reinforced concrete section at ULS is solicited in combined bending and axial compression. The concrete constitutive law at ULS is the parabola–rectangle constitutive law. The section is composed of tensile and compression steel reinforcements (see Figure 1.1). The steel reinforcements are assumed to behave in the plasticity range with linear hardening. The problem consists of designing the tensile steel reinforcement of the reinforced concrete section at the ULS, for a given bending moment and normal force solicitation. The geometrical parameters of the problem are given by:

$$b=0.2 \text{ m}; d=0.45 \text{ m}; d'=0.05 \text{ m}; h=0.50 \text{ m};$$

$$A_{s2} =0.848 \text{ cm}^2; A_{s1}: \text{unknown} \qquad [1.131]$$

The compression and tensile steel reinforcements are made up of steel B500B steel bars. The hardening effect is taken into account. The steel parameters at ULS are given by equations [1.107] and [1.108]. The concrete has a class of C35/45 and is modeled with a parabola–rectangle constitutive law, characterized by the characteristics values:

$$\varepsilon_{c2} = - 2 \text{ \textperthousand}; \ \varepsilon_{cu2} = - 3,5 \text{ \textperthousand}; \ f_{cu} = -\alpha_{cc} \frac{f_{ck}}{\gamma_c} = -21 \text{ MPa}$$

$$\text{with } \alpha_{cc} = 0.9 \qquad [1.132]$$

The coefficient α_{cc} has been chosen equal to $\alpha_{cc} = 0.9$ for this design problem. The solicitation is given at the point S of Figure 1.14, that is at the upper fiber of the reinforced cross-section:

$$M_S = 0.099 \text{ MN.m} \quad \text{and} \quad N_S = -0.18 \text{ MN} \qquad [1.133]$$

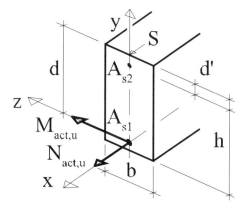

Figure 1.14. *Definition of the reinforced cross-section solicited by combined bending with normal forces*

The exercise consists of designing the tensile steel reinforcement area of this reinforced concrete section at the ULS for the given solicitation in combined bending and normal force (N_S, M_S). The solicitation is equivalently calculated at the center of gravity of the tensile steel reinforcement as:

$$M_u = M_S - N_S d = 0.099 + 0.45 \times 0.18$$
$$= 0.18 \text{ MN.m and } N_u = N_S = -0.18 \text{ MN} \qquad [1.134]$$

This design needs to determine in an analytical and numerical way the sectional resistant moment at the ULS calculated at the center of gravity of the tensile steel reinforcement $M_{res,u}(\alpha)$. This resistant moment is a function of the dimensionless relative depth α. This function is continuous and piecewise differentiable. For the parameter of interest, the resolution of the nonlinear function $M_{res,u}(\alpha) = M_u$ at the center of gravity of the tensile steel reinforcement, gives the parameter α associated with the working behavior of the cross-section, and then the tensile steel area can be calculated from the application of the normal force equilibrium equation.

1.2.2.2. Resolution – pivot A1

Pivot $A1$ is characterized by the domain of variation in the position of the neutral axis position:

$$\alpha \in [0; \alpha_{A_1 A_2}] \text{ with } \alpha_{A_1 A_2} = \frac{-\varepsilon_{c2}}{\varepsilon_{ud} - \varepsilon_{c2}} = \frac{2}{47} = 0.042553 \qquad [1.135]$$

The section behaves in pivot A meaning that the tensile steel reinforcements have reached their strain limit capacity $\varepsilon_{s1} = \varepsilon_{ud}$, and the concrete behaves in the parabolic regime $\varepsilon_{c,sup} \in [\varepsilon_{c2}; 0]$. It can be shown, for this range of parameters, that the upper steel reinforcement are in the plasticity regime:

$$\varepsilon_{s2}(\alpha = 0) = \delta' \varepsilon_{ud} = 5\text{\textperthousand} > \varepsilon_{su} = 2.17\text{\textperthousand} \text{ and } \varepsilon_{s2}(\alpha = \alpha_{A_1 A_2})$$

$$= \frac{\alpha_{A_1 A_2} - \delta'}{\alpha_{A_1 A_2}} \varepsilon_{c2} = 3.222222\text{\textperthousand} > \varepsilon_{su} = 2.17\text{\textperthousand} \qquad [1.136]$$

In pivot A_1, the reinforcement in the upper part of the cross-section works in tension in the plasticity regime:

$$\sigma_{s2} = q\varepsilon_{s2} + q' \geq 0 \qquad [1.137]$$

Using equation [3.145] of [CAS 12], the resistant moment calculated at the center of gravity of the tensile steel reinforcement is given for $\alpha \in [0; \alpha_{A_1 A_2}]$ as:

$$M_{res,u}(\alpha) = -bd^2 f_{cu} \frac{\alpha^2 \varepsilon_{ud}}{12(1-\alpha)^2 \varepsilon_{c2}^2} \times$$
$$(-12\varepsilon_{c2} + 16\alpha\varepsilon_{c2} - 4\alpha^2 \varepsilon_{c2} - 4\alpha\varepsilon_{ud} + \alpha^2 \varepsilon_{ud})$$
$$- A_{s2}(d-d') \left[q \frac{\alpha d - d'}{(\alpha-1)d} \varepsilon_{ud} + q' \right] \qquad [1.138]$$

We note, in this case, that the equation $M_{res,u}(\alpha) = M_u$ is a quartic equation to be solved for calculating the unknown position of the neutral axis α.

1.2.2.3. Resolution – pivot A2

Pivot $A2$ is characterized by the domain of variation in the position of the neutral axis position:

$$\alpha \in [\alpha_{A_1 A_2}; \alpha_{AB}] \text{ with } \alpha_{A_1 A_2} = \frac{-\varepsilon_{c2}}{\varepsilon_{ud} - \varepsilon_{c2}} = \frac{2}{47} = 0.042553 \text{ and}$$

$$\alpha_{AB} = \frac{-\varepsilon_{cu2}}{\varepsilon_{ud} - \varepsilon_{cu2}} = \frac{3.5}{48.5} = 0.072165 \qquad [1.139]$$

The section behaves in pivot A, which means that the tensile steel reinforcements have reached their strain limit capacity $\varepsilon_{s1} = \varepsilon_{ud}$, and the concrete behaves in the elastoplastic regime $\varepsilon_{c,sup} \in [\varepsilon_{cu2}; \varepsilon_{c2}]$. It can be shown for this range of parameters that the upper steel reinforcement is still in the tensile regime, as the strain in the upper steel reinforcement, at each bound of the interval $\alpha \in [\alpha_{A_1 A_2}; \alpha_{AB}]$ is an extension strain:

$$\varepsilon_{s2}(\alpha = \alpha_{A_1 A_2}) = \frac{\alpha_{A_1 A_2} - \delta'}{\alpha_{A_1 A_2}} \varepsilon_{c2} = 3.222222\text{‰} > \varepsilon_{su} = 2.17\text{‰} \text{ and}$$

$$\varepsilon_{s2}(\alpha = \alpha_{AB}) = \frac{\alpha_{AB} - \delta'}{\alpha_{AB}} \varepsilon_{cu2} = 1.888888\text{‰} < \varepsilon_{su} = 2.17\text{‰} \qquad [1.140]$$

In pivot A_2, the reinforcement in the upper part of the cross-section works in tension:

$$\sigma_{s2} = m\varepsilon_{s2} + m' \geq 0 \qquad [1.141]$$

where the parameters m and m' depend on the value of α as:

$$m = q \text{ and } m' = q' \text{ for } \alpha \in \left[\alpha_{A_1 A_2}; \frac{\delta' \varepsilon_{ud} - \varepsilon_{su}}{\varepsilon_{ud} - \varepsilon_{su}} \right] = [0.042553; 0.065990]$$

$$m = E_s \text{ and } m' = 0 \text{ for } \alpha \in \left[\frac{\delta' \varepsilon_{ud} - \varepsilon_{su}}{\varepsilon_{ud} - \varepsilon_{su}}; \alpha_{AB} \right] = [0.065990; 0.072165] \qquad [1.142]$$

More generally, the upper steel reinforcements are always in the tensile regime in pivot A for:

$$\delta' = \frac{d'}{d} \geq \alpha_{AB} \qquad [1.143]$$

It can be checked that this last inequality is valid in the reinforced concrete section studied here, as $\delta' = 0.1111 \geq \alpha_{AB} = 0.0722$. Using equation [3.149] of [CAS 12], the resistant moment calculated at the center of gravity of the tensile steel reinforcement is given for $\alpha \in [\alpha_{A_1 A_2} \cdot \alpha_{AB}]$ as:

$$M_{res,u}(\alpha) = -bd^2 f_{cu} \frac{\varepsilon_{c2}(4\varepsilon_{ud} - \varepsilon_{c2})(1-\alpha)^2 + 6\alpha\varepsilon_{ud}^2(2-\alpha)}{12\varepsilon_{ud}^2}$$

$$-A_{s2}(d - d')\left[m \frac{\alpha d - d'}{(\alpha - 1)d} \varepsilon_{ud} + m' \right] \qquad [1.144]$$

We note, in this case, that the equation $M_{res,u}(\alpha) = M_u$ is a cubic equation to be solved for calculating the unknown position of the neutral axis α.

1.2.2.4. *Resolution – pivot B*

Pivot B is characterized by the domain of variation for the position of the neutral axis:

$$\alpha \in [\alpha_{AB};1] \text{ with } \alpha_{AB} = \frac{-\varepsilon_{cu2}}{\varepsilon_{ud} - \varepsilon_{cu2}} = \frac{3.5}{48.5} = 0.072165 \qquad [1.145]$$

The compression concrete block has reached its ultimate strain capacity characterized by $\varepsilon_{c,\sup} = \varepsilon_{cu2}$. Using equation [3.154] of [CAS 12], the resistant moment calculated at the center of gravity of the tensile steel reinforcement is given for $\alpha \in [\alpha_{AB};1]$ as:

$$M_{res,u}(\alpha) = -bd^2 f_{cu} \left[\frac{-\alpha^2}{12} \left(\frac{\varepsilon_{c2}}{\varepsilon_{cu2}} \right)^2 + \alpha \left(1 - \frac{\alpha}{2} \right) + \frac{\alpha}{3} \left(\frac{\varepsilon_{c2}}{\varepsilon_{cu2}} \right) (\alpha - 1) \right]$$
$$- A_{s2}(d - d') \left[m \frac{\alpha d - d'}{\alpha d} \varepsilon_{cu2} + m' \right] \qquad [1.146]$$

We also note, in this case, that the equation $M_{res,u}(\alpha) = M_u$ is a cubic equation to be solved for calculating the unknown position of the neutral axis α.

The value of the parameters m and m' depends on the dimensionless position of neutral axis α as:

$$m = E_s \text{ and } m' = 0 \text{ for } \alpha \in \left[\alpha_{AB}; \frac{\delta' \varepsilon_{cu2}}{\varepsilon_{cu2} + \varepsilon_{su}} \right] = [0.072165; 0.293260]$$

$$m = q \text{ and } m' = -q' \text{ for } \alpha \in \left[\frac{\delta' \varepsilon_{cu2}}{\varepsilon_{cu2} + \varepsilon_{su}}; 1 \right] = [0.293260; 1] \qquad [1.147]$$

We note that the compression steel reinforcements are in the plasticity branch in pivot B for:

$$\frac{\delta' \varepsilon_{cu2}}{\varepsilon_{cu2} + \varepsilon_{su}} \geq \alpha_{AB} \Rightarrow \delta' \geq \frac{\varepsilon_{su} + \varepsilon_{cu2}}{\varepsilon_{cu2} - \varepsilon_{ud}} = 0.027342 \qquad [1.148]$$

For the considered reinforced concrete section with $\delta' = 0.11111$, this inequality holds and it is confirmed that the compression steel reinforcement reaches the yield stress in compression in pivot B.

1.2.2.5. Synthesis – nonlinear problem to be solved

We calculate the moment at the transition between pivot A and pivot B from equation [1.146] detailed below with $m = E_s$ and $m' = 0$:

$$M_{AB} = M_{res,u}(\alpha = \alpha_{AB})$$

$$= -bd^2 f_{cu} \left[\frac{-\alpha_{AB}^2}{12}\left(\frac{\varepsilon_{c2}}{\varepsilon_{cu2}}\right)^2 + \alpha_{AB}\left(1 - \frac{\alpha_{AB}}{2}\right) + \frac{\alpha_{AB}}{3}\left(\frac{\varepsilon_{c2}}{\varepsilon_{cu2}}\right)(\alpha_{AB} - 1) \right]$$

$$- A_{s2}(d-d')E_s \frac{\alpha_{AB} - \delta'}{\alpha_{AB}}\varepsilon_{cu2} = 0.035380 \text{ MN.m} \qquad [1.149]$$

It can be seen that $M_u = 0.18$ MN.m $> M_{AB} = 0.035380$ MN.m and the section behaves in pivot B. Moreover, it can be shown, according to Figure 1.15, that:

$$M_u = 0.14 \text{MN.m} \in \left[M_{res,u}(\alpha = \alpha_{AB}); M_{res,u}\left(\alpha = \frac{\delta' \varepsilon_{cu3}}{\varepsilon_{cu3} + \varepsilon_{su}}\right) \right] \qquad [1.150]$$

The section behaves in pivot B with compression steel reinforcement in elasticity for $\mu_{res,u}(\alpha) = \mu_u = 0.211640$. The dimensionless position of the neutral axis has then to be calculated from the nonlinear equation [1.146] specialized with the compression steel reinforcement in the elasticity range, and written as:

$$M_u = -bd^2 f_{cu} \left[\frac{-\alpha^2}{12}\left(\frac{\varepsilon_{c2}}{\varepsilon_{cu2}}\right)^2 + \alpha\left(1 - \frac{\alpha}{2}\right) + \frac{\alpha}{3}\left(\frac{\varepsilon_{c2}}{\varepsilon_{cu2}}\right)(\alpha - 1) \right]$$

$$- A_{s2}(d-d')E_s \frac{\alpha - \delta'}{\alpha}\varepsilon_{cu2} \qquad [1.151]$$

which can be written as a cubic equation:

$$\alpha^3 \left[\frac{1}{12}\left(\frac{\varepsilon_{c2}}{\varepsilon_{cu2}}\right)^2 + \frac{1}{2} - \frac{1}{3}\left(\frac{\varepsilon_{c2}}{\varepsilon_{cu2}}\right) \right] + \alpha^2 \left[-1 + \frac{1}{3}\left(\frac{\varepsilon_{c2}}{\varepsilon_{cu2}}\right) \right]$$

$$+ \alpha \left[-\frac{E_s A_{s2}}{bd f_{cu}}(1-\delta')\varepsilon_{cu2} - \frac{M_u}{bd^2 f_{cu}} \right]$$

$$+ \frac{E_s A_{s2}}{bd f_{cu}}\delta'(1-\delta')\varepsilon_{cu2} = 0 \qquad [1.152]$$

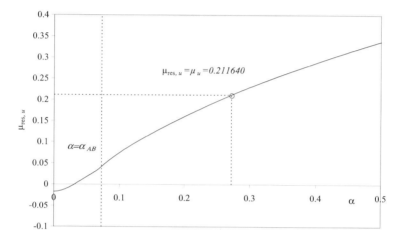

Figure 1.15. *Evolution of the reduced moment $\mu_{res,u}$ with respect to α; graphical resolution for $\mu_{res,u}(\alpha)=\mu_u=0.211640$; $\alpha=0.271775$*

The cubic equation is numerically described as:

$$a\alpha^3 + b\alpha^2 + c\alpha + d = 0 \text{ with } a = 49.5, b = -119,$$
$$c = 27.00721 \text{ and } d = 0.455989 \qquad [1.153]$$

Cardano's method can be used as detailed in Appendix 1. The canonical parameters p and q are calculated as:

$$p = \frac{3ac - b^2}{3a^2} = -1.380870 \text{ and}$$

$$q = \frac{27a^2 d + 2b^3 - 9abc}{27a^3} = -0.582754 \qquad [1.154]$$

leading to $4p^3 + 27q^2 = -2.18488 < 0$. The cubic has three real roots calculated as:

$$
\begin{cases}
\alpha_1 = 2\sqrt{\dfrac{-p}{3}}\cos\left[\dfrac{Arc\cos\left(\dfrac{3q}{2p}\sqrt{\dfrac{3}{-p}}\right)+2\pi}{3}\right] - \dfrac{b}{3a} = -0.015780 \\[4ex]
\alpha_2 = 2\sqrt{\dfrac{-p}{3}}\cos\left[\dfrac{Arc\cos\left(\dfrac{3q}{2p}\sqrt{\dfrac{3}{-p}}\right)+4\pi}{3}\right] - \dfrac{b}{3a} = 0.271775 \\[4ex]
\alpha_3 = 2\sqrt{\dfrac{-p}{3}}\cos\left[\dfrac{Arc\cos\left(\dfrac{3q}{2p}\sqrt{\dfrac{3}{-p}}\right)}{3}\right] - \dfrac{b}{3a} = 2.148046
\end{cases}
\quad [1.155]
$$

The solution of the problem is $\alpha = \alpha_2 = 0.271775$, the other two solutions being negative or greater than unity. For this value of the neutral position $\alpha = \alpha_2 = 0.271775$, the strain in the tensile and compression steel reinforcements can be calculated as:

$$\varepsilon_{s1} = \frac{\alpha - 1}{\alpha}\varepsilon_{cu2} = 9.378\text{\textperthousand} \gg \varepsilon_{su} = 2.17\text{\textperthousand} \text{ and}$$

$$\varepsilon_{s2} = \frac{\alpha - \delta'}{\alpha}\varepsilon_{cu2} = -2.069\text{\textperthousand} > -\varepsilon_{su} = -2.17\text{\textperthousand} \qquad [1.156]$$

It is easily checked that the tensile steel reinforcements work in plasticity whereas the compression steel reinforcements are in the elastic range. The steel reinforcement area is then computed from the normal force equilibrium equation (see also equation [3.155] of [CAS 12] for the component of the compression concrete block):

$$A_{s1}\sigma_{s1} + A_{s2}\sigma_{s2} + bdf_{cu}\alpha\psi_B = N_u \text{ with } \psi_B = \frac{17}{21} \qquad [1.157]$$

The tensile steel reinforcement area is then expressed from:

$$A_{s1} = \frac{N_u - A_{s2}E_s\varepsilon_{s2} - \alpha bd\psi_B f_{cu}}{\sigma_{s1}} \text{ with}$$

$$\sigma_{s1} = q\varepsilon_{s1} + q' = 440.023 \text{ MPa} \tag{1.158}$$

We finally obtain $A_{s1} = 6.157 \text{ cm}^2$. The section can be designed with $3\phi20$ for the tensile steel reinforcement ($A_{s1} = 9.425 \text{ cm}^2$).

1.2.3. T-cross-section – general resolution for bilinear or parabola–rectangle laws for concrete

1.2.3.1. The T-cross-section modeled with two rectangular cross-sections

In this section, the design at ULS of the reinforced T-cross-section with both tensile (with area A_{s1}) and compression (with area A_{s2}) steel reinforcements is analyzed. The concrete is modeled by a bilinear or a parabola–rectangle constitutive law. The steel reinforcements are assumed to behave either with an elastic and perfectly plastic constitutive law, or with the possible linear plastic hardening. The geometry of the cross-section is characterized by the different length parameters b, b_w, h, h_0, d, d'. b is the width of the concrete slab, h_0 is the depth of the flange (slab) thickness. The width of the web is denoted by b_w. The position of the neutral axis is, as usual, characterized by $y = \alpha d$ from the upper fiber of the cross-section. The reinforced concrete cross-section is shown in Figure 1.16.

For a general reinforced concrete section, the bending moment and normal force equilibrium equations are written with respect to the center of gravity of the tensile steel reinforcement, as:

$$\begin{cases} M_u = M_c - A_{s2}\sigma_{s2}(d - d') \\ N_u = N_c + \sigma_{s1}A_{s1} + \sigma_{s2}A_{s2} \end{cases} \text{ with } y = \alpha d \tag{1.159}$$

The calculation of the internal actions in the compression concrete block depends on the position of the neutral axis. If the depth of the compression block is within the flanged portion of the beam, that is the size of the compression block αd is less than the flange (slab) thickness h_0, measured from the top of the slab ($y = \alpha d < h_0$), then the section can be calculated as an "equivalent" rectangular cross-section with the width equal to b (as tensile concrete contribution is neglected in the analysis). Then, we find again the configuration previously investigated, for the design of reinforced concrete beams with the rectangular cross-section at the ULS (see Figure 1.16). However, when the size of the compression block is larger than the flange (slab) thickness ($y = \alpha d > h_0$), the calculation has to be based on the T-cross-section calculation.

The transition from a rectangular-type cross-section calculation and a T-type cross-section calculation corresponds to the total compression of the flanged portion of the cross-section, characterized by $y = \alpha d = h_0$.

The distinction between the two types of calculation (rectangular cross-section calculation or T-cross-section calculation) can be made from the following criterion:

$$y = \alpha d \leq h_0 \rightarrow \text{Calculation based on a rectangular cross-section}$$
$$y = \alpha d \geq h_0 \rightarrow \text{Calculation based on a T-cross-section} \qquad [1.160]$$

We note that these conditions are slightly different from the conditions used for the simplified rectangular concrete law – see equation [1.58].

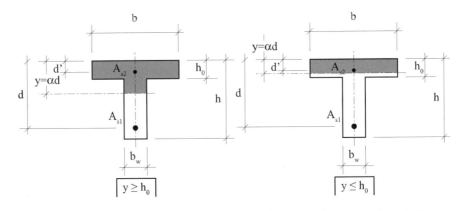

Figure 1.16. *Position of the neutral axis at ultimate limit state; T-cross-section – bilinear or parabolic-rectangle constitutive law for concrete*

In the following, the case of a T-cross-section calculation is detailed, as one on the rectangular cross-section has already been investigated. When the size of the compression block is larger than the flange (slab) thickness

($y = \alpha d > h_0$), the design of the T-cross-section can be decomposed into two rectangular cross-sections, as shown in Figure 1.17.

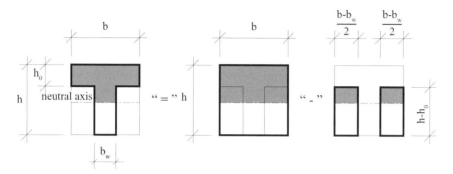

Figure 1.17. *A T-cross-section modeled with two rectangular cross-sections*

The T-cross-section can be decomposed into two fictitious cross-sections: the first cross-section is a rectangular cross-section with a width b, effective height d and neutral axis position parameter α, and the second cross-section is also rectangular with a width b', effective height d'' and neutral axis position parameter α' given as:

$$b' = b - b_w; d'' = d - h_0 \text{ and } \alpha' = \frac{\alpha d - h_0}{d - h_0} \qquad [1.161]$$

In particular, we check that $\alpha' d'' = \alpha d - h_0$. The decomposition of the T-cross-section into two rectangular cross-sections is used for the internal screw elements (N_c, M_c) in the concrete block, calculated at the center of gravity of the tensile steel reinforcements. The internal components in the compression block can then be calculated as:

$$\begin{cases} M_c = -bd^2 f_{cu}\mu + b'd''^2 f_{cu}\mu' \\ N_c = \alpha\psi bdf_{cu} - \alpha'\psi'b'd'f_{cu} \end{cases} \qquad [1.162]$$

The equilibrium equations are then written as:

$$\begin{cases} M_u = -bd^2 f_{cu}\mu + (b - b_w)(d - h_0)^2 f_{cu}\mu' - A_{s2}\sigma_{s2}(d - d') \\ N_u = \alpha\psi bdf_{cu} - (\alpha d - h_0)(b - b_w)\psi' f_{cu} + \sigma_{s1}A_{s1} + \sigma_{s2}A_{s2} \end{cases} \qquad [1.163]$$

The T-cross-section can be analyzed as an equivalent rectangular reinforced concrete section through an equivalent reduced moment μ_T defined by:

$$\mu_T = \frac{M_c}{-bd^2 f_{cu}} = \mu - \left(1 - \frac{b_w}{b}\right)\left(1 - \frac{h_0}{d}\right)^2 \mu' \qquad [1.164]$$

In the following, the reduced moments μ and μ' are detailed for each pivot. This methodology is detailed by Casandjian [CAS 89], and the reduced parameters are given also in [CAS 89] for a concrete model based on a parabolic–rectangle law.

1.2.3.2. Pivot A – T-cross-section calculation

In pivot $A1$, the reduced parameters are calculated from the reduced parameters identified in pivot $A1$ for the rectangular cross-section as:

$$\begin{cases} \mu = \mu_{A1} \\ \mu' = \mu'_{A1} \end{cases} \text{and} \begin{cases} \alpha\psi = \alpha\psi_{A1} \\ \alpha'\psi' = \alpha'\psi'_{A1} \end{cases} \qquad [1.165]$$

where the prime parameters are defined in equation [1.161]. For instance, in pivot $A1$ for the bilinear concrete law, we simply calculate from equations [3.121] and [3.122] of [CAS 12]:

$$\begin{cases} \mu = \mu_{A1} = \dfrac{\alpha^2}{2(1-\alpha)}\left(1 - \dfrac{\alpha}{3}\right)\left(\dfrac{\varepsilon_{ud}}{-\varepsilon_{c3}}\right) \\ \\ \mu' = \mu'_{A1} = \dfrac{\alpha'^2}{2(1-\alpha')}\left(1 - \dfrac{\alpha'}{3}\right)\left(\dfrac{\varepsilon_{ud}}{-\varepsilon_{c3}}\right) \end{cases} \text{and}$$

$$\begin{cases} \alpha\psi = \alpha\psi_{A1} = \dfrac{\alpha^2}{2(1-\alpha)}\left(\dfrac{\varepsilon_{ud}}{-\varepsilon_{c3}}\right) \\ \\ \alpha'\psi' = \alpha'\psi'_{A1} = \dfrac{\alpha'^2}{2(1-\alpha')}\left(\dfrac{\varepsilon_{ud}}{-\varepsilon_{c3}}\right) \end{cases} \qquad [1.166a]$$

As another example, in pivot $A1$ for the parabola–rectangle concrete law, we calculate from equations [3.144] and [3.145] of [CAS 12]:

$$\begin{cases} \mu = \mu_{A1} = \dfrac{\alpha^2 \varepsilon_{ud}(-12\varepsilon_{c2} + 16\alpha\varepsilon_{c2} - 4\alpha^2\varepsilon_{c2} - 4\alpha\varepsilon_{ud} + \alpha^2\varepsilon_{ud})}{12(1-\alpha)^2 \varepsilon_{c2}^2} \\[4mm] \mu' = \mu'_{A1} = \dfrac{\alpha'^2 \varepsilon_{ud}(-12\varepsilon_{c2} + 16\alpha'\varepsilon_{c2} - 4\alpha'^2\varepsilon_{c2} - 4\alpha'\varepsilon_{ud} + \alpha'^2\varepsilon_{ud})}{12(1-\alpha')^2 \varepsilon_{c2}^2} \end{cases}$$

and

$$\begin{cases} \alpha\psi = \alpha\psi_{A1} = \dfrac{\alpha^2 \varepsilon_{ud}(-3\varepsilon_{c2} - \alpha\varepsilon_{ud} + 3\alpha\varepsilon_{c2})}{3(1-\alpha)^2 \varepsilon_{c2}^2} \\[4mm] \alpha'\psi' = \alpha'\psi'_{A1} = \dfrac{\alpha'^2 \varepsilon_{ud}(-3\varepsilon_{c2} - \alpha'\varepsilon_{ud} + 3\alpha'\varepsilon_{c2})}{3(1-\alpha')^2 \varepsilon_{c2}^2} \end{cases} \qquad [1.166b]$$

In pivot $A2$, two cases can be distinguished depending on the position of the junction of the two parts of the bilinear or the parabola–rectangle concrete law, with respect to the flange junction:

$$\alpha_c d \le h_0 \implies \begin{cases} \mu = \mu_{A2} \\ \mu' = \mu'_{A1} \end{cases} \text{and} \begin{cases} \alpha\psi = \alpha\psi_{A2} \\ \alpha'\psi' = \alpha'\psi'_{A1} \end{cases}$$

$$\alpha_c d \ge h_0 \implies \begin{cases} \mu = \mu_{A2} \\ \mu' = \mu'_{A2} \end{cases} \text{and} \begin{cases} \alpha\psi = \alpha\psi_{A2} \\ \alpha'\psi' = \alpha'\psi'_{A2} \end{cases}$$

$$\text{with } \alpha_c = \alpha - (1-\alpha)\left(\dfrac{-\varepsilon_{ci}}{\varepsilon_{ud}}\right) \qquad [1.167]$$

where $\varepsilon_{ci} = \varepsilon_{c3}$ for the bilinear concrete law, and $\varepsilon_{ci} = \varepsilon_{c2}$ for the parabola–rectangle concrete law.

In both pivot $A1$ and pivot $A2$, we have for the calculation of the stress in the compression steel reinforcement:

$$\sigma_{s2} = -m\varepsilon_{ud}\dfrac{\alpha - \delta'}{1-\alpha} + m' \qquad [1.168]$$

1.2.3.3. Pivot B – T-cross-section calculation

We now detail the design of the T-cross-section in pivot B for the bilinear rectangular constitutive law for concrete. The results for the parabola–rectangle concrete law can be found in [CAS 89].

In pivot B, for the bilinear rectangular constitutive law for concrete, equations [3.132] and [3.133] of [CAS 12] give for $\varepsilon_{cu3} = 2\varepsilon_{c3}$:

$$\mu_B = \frac{3}{4}\alpha\left(1 - \frac{7}{18}\alpha\right) \text{ and } \alpha\psi_B = \frac{3}{4}\alpha \qquad [1.169]$$

The calculation of the critical parameter α_c associated with the junction between the elastic and the constant stress part is given in pivot B by:

$$\alpha_c = \alpha\left(1 - \frac{\varepsilon_{c3}}{\varepsilon_{cu3}}\right) = \frac{\alpha}{2} \qquad [1.170]$$

The determination of μ' and ψ' depends on the position of the junction of the two parts of the bilinear law with respect to the flange junction:

$\alpha_c d \leq h_0 \rightarrow$ Calculation of μ' with elastic behaviour of concrete

$\alpha_c d \geq h_0 \rightarrow$ Calculation of μ' with elasto-plastic behaviour of concrete   [1.171]

Assuming first $\alpha_c d \geq h_0$, the internal bending moment in the concrete block of the fictitious cross-section can be calculated from:

$$M'_c = -b'd''f_{cu}\alpha'_c\left(d'' - \frac{\alpha'_c d''}{2}\right) - \frac{b'd''f_{cu}}{2}(\alpha' - \alpha'_c)$$

$$\left[\frac{2d''}{3}(\alpha' - \alpha'_c) + (1 - \alpha')d''\right] \text{with } \alpha'_c = \frac{\alpha_c d - h_0}{d - h_0} \qquad [1.172]$$

When considering $\alpha_c d \leq h_0$, the internal bending moment in the concrete block is calculated for the elastic stress field as:

$$M'_c = \frac{-\alpha'b'd''\sigma'_{c,\text{sup}}}{2}\left[(1 - \alpha')d'' + \frac{2}{3}\alpha'd''\right] \text{ with}$$

$$\sigma'_{c,\text{sup}} = f_{cu}\frac{\alpha'd''}{\alpha d}\frac{\varepsilon_{cu3}}{\varepsilon_{c3}} \qquad [1.173]$$

The calculations are now detailed in this second case, associated with the reduced parameters of the second fictitious rectangular cross-section:

$$\mu' = \frac{M'_c}{-b'd''^2 f_{cu}} = \frac{\alpha'}{2}\left(1 - \frac{\alpha'}{3}\right)\frac{\varepsilon_{cu3}}{\varepsilon_{c3}}\frac{\alpha'd''}{\alpha d} \quad \text{and} \quad \alpha'\psi' = \frac{\alpha'}{2}\frac{\alpha'd''}{\alpha d}\frac{\varepsilon_{cu3}}{\varepsilon_{c3}} \quad [1.174]$$

In pivot B, we have for the calculation of the stress in the compression steel reinforcement:

$$\sigma_{s2} = m\varepsilon_{cu3}\frac{\alpha - \delta'}{\alpha} + m' \tag{1.175}$$

Inserting equations [1.174] and [1.175] in [1.163] leads to a cubic equation for the determination of the dimensionless position of the neutral axis α:

$$(\alpha d)^3\left[-f_{cu}\left(\frac{b}{24} - \frac{b_w}{3}\right)\right] + (\alpha d)^2\left[-df_{cu}\left(-\frac{b}{4} + b_w\right)\right]$$
$$+ (\alpha d)\left[-f_{cu}h_0(b - b_w)(2d - h_0) - M_u - A_{s2}(d - d')(m' + m\varepsilon_{cu3})\right]$$
$$+ \frac{(b - b_w)(-f_{cu})}{3}h_0^2(2h_0 - 3d) + A_{s2}(d - d')dm\varepsilon_{cu3}\delta' = 0 \tag{1.176}$$

For a reinforced concrete section without compression steel reinforcement, this cubic equation is reduced to:

$$(\alpha d)^3\left[-f_{cu}\left(\frac{b}{24} - \frac{b_w}{3}\right)\right] - (\alpha d)^2\left[-df_{cu}\left(\frac{b}{4} - b_w\right)\right] +$$
$$(\alpha d)[-f_{cu}h_0(b - b_w)(2d - h_0) - M_u] - \frac{(b - b_w)(-f_{cu})}{3}h_0^2(3d - 2h_0) = 0 \tag{1.177}$$

which is also coincident with the results presented in [SIE 10].

The normal force in the concrete block is given by:

$$N_c = \frac{3}{4}\alpha bdf_{cu} - (b - b_w)f_{cu}\frac{(\alpha d - h_0)^2}{\alpha d} \tag{1.178}$$

The tensile steel reinforcement area is then computed from:

$$A_{s1} = \frac{N_u - A_{s2}\sigma_{s2} + \frac{3}{4}\alpha bd(-f_{cu}) - (b-b_w)(-f_{cu})\dfrac{(\alpha d - h_0)^2}{\alpha d}}{\sigma_{s1}} \qquad [1.179]$$

For a reinforced concrete T-cross-section with only tensile steel reinforcement ($A_{s2} = 0$) and with simple bending solicitation ($N_u = 0$), the tensile steel area A_{s1} is calculated from:

$$A_{s1} = \frac{\frac{3}{4}\alpha bd(-f_{cu}) - (b-b_w)(-f_{cu})\dfrac{(\alpha d - h_0)^2}{\alpha d}}{f_{su}} \qquad [1.180]$$

where the hardening effect is neglected, as a simplification, and it is assumed that the balanced failure point has not been reached. If the hardening effect in the reinforcement bars is taken into account, the stress in the tensile steel reinforcement has to be explicitly calculated.

1.2.3.4. *Design of a reinforced concrete T-cross-section – example*

1.2.3.4.1. Data of the problem

A T-cross-section is analyzed under simple bending (without additional normal force *Nact* = 0), with tensile steel reinforcement (with area A_{s1}), in the lower part of the cross-section (see Figure 1.18). This section corresponds to the central zone of an isostatic beam (length L = 6.00 m) under a concrete slab of thickness 20 cm.

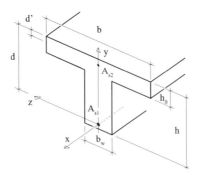

Figure 1.18. *Geometry of a T-cross-section with only tensile steel reinforcements*

The geometrical parameters of the problem are given by:

$$b = 2.28 \text{ m}; b_w = 0.2 \text{ m}; d = 0.9h = 0.405 \text{ m}; \ h = 0.45 \text{ m};$$
$$h_0 = 0.2 \text{ m}; \ A_{s1}: \text{unknown} \qquad [1.181]$$

The compression and tensile steel reinforcements are made up of steel B500B steel bars. Hardening effect is taken into account. The steel parameters at ULS are:

$E_s = 200$ GPa; $\varepsilon_{ud} = 45‰$; $\varepsilon_{uk} = 50‰$; $f_{su} = 434.783$ MPa; $k = 1.08$ leading to:

$$q = \frac{(k-1)f_{su}}{\varepsilon_{uk} - \varepsilon_{su}} = 727.273 \text{ MPa},$$

$$q' = f_{su}\left(1 - \frac{q}{E_s}\right) = 433.202 \text{ MPa}$$

$$\text{and } \varepsilon_{su} = \frac{f_{su}}{E_s} = 2.17‰ \qquad [1.182]$$

The concrete has a class $C25/30$-type and is modeled with a bilinear constitutive law, characterized by the characteristics values as $f_{ck} \leq 50$ MPa:

$$\varepsilon_{c3} = -1.75‰; \varepsilon_{cu3} = -3.5‰;$$

$$f_{cu} = -\alpha_{cc}\frac{f_{ck}}{\gamma_c} = -16.67 \text{ MPa with } \alpha_{cc} = 1 \text{ and } \gamma_c = 1.5 \qquad [1.183]$$

According to the French National Annexes, the coefficient α_{cc} has been chosen equal to $\alpha_{cc} = 1$ for this design problem.

The steel cover C depends on the environmental class (see Eurocode 2). In Eurocode 2, the nominal steel cover C_{nom} is given by the following formula:

$$C_{nom} = C_{min} + \Delta C_{dev} \text{ with } C_{min} = \max{(\Phi, C_{min,dur}, 10 \text{ mm})} \qquad [1.184]$$

where Φ denotes the bars diameter. In the present design, the environmental class is XC1, and for on-site casting beam and slab, we use the values

ΔC_{dev} = 10 mm (on-site casting) and $C_{min,dur}$ = 15 mm (XC1 class), finally leading to the nominal steel cover:

$$C_{nom} = \max (25 \text{ mm} ; \Phi + 10 \text{ mm}) \qquad [1.185]$$

The bending moment of permanent loading on the cross-section is M_G = 130 kN.m, and the bending moment of variable loading on the cross-section is M_Q = 31 kN.m. The ULS solicitation is then calculated as:

$$M_u = M_{Ed} = 1.35 M_G + 1.5 M_Q = 222 \text{ kN.m} \qquad [1.186]$$

With such concrete and steel behaviors, the change from pivot A to pivot B corresponds to (see equation [3.115] of [CAS 12]):

$$\alpha_{AB} = -\varepsilon_{cu3} / (\varepsilon_{ud} - \varepsilon_{cu3}) = 0.0721649 \qquad [1.187]$$

The change from pivot $A1$ to pivot $A2$ corresponds to (see equation [3.119] of [CAS 13]):

$$\alpha_{A1A2} = -\varepsilon_{c3} / (\varepsilon_{ud} - \varepsilon_{c3}) = 0.037433 \qquad [1.188]$$

1.2.3.4.2. Rectangular cross-section behavior – M_u=222 kN.m

If the depth of the compression zone y is less than h_0, the T-cross-section behaves as a rectangular section with a width b. If $y > h_0$, a part of the web is under compression. The simple bending moment obtained with $y = h_0$ is denoted by M_T (see equation [1.59]). If M_T would have been calculated with the pivot A assumption, then the plane cross-section assumption would have induced:

$$\varepsilon_{c,sup} = - \varepsilon_{ud} \, h_0/(d - h_0) = - 43.9 \text{ \textperthousand} \leq -3.5\text{\textperthousand} \qquad [1.189]$$

which is a non-acceptable value for the concrete compressive strain. It means that M_T has to be calculated with the assumption of pivot B, and the tensile steel strain in this case is, in fact, acceptable as:

$$\varepsilon_{ud} = -\varepsilon_{cu3}(d - h_0)/ h_0 = 3.59\text{\textperthousand} \leq 45\text{\textperthousand} \qquad [1.190]$$

The value of M_T is calculated from:

$$M_T = -f_{cu}bh_0(d - h_0/4)/2 - f_{cu}bh_0(d - 2h_0/3)/4$$
$$= 1865.5 \text{ kN.m} \tag{1.191}$$

M_T is higher than M_u. Then, the neutral axis is within the flange thickness $(y = \alpha d \leq h_0)$. The T-cross-section can be calculated as a rectangular one with a width b. The reduced moment is obtained from:

$$\mu_u = -M_u/(bd^2f_{cu}) = 0.03561 \tag{1.192}$$

The position of the neutral axis α is calculated in the case of this equivalent rectangular cross-section without compression steel reinforcement from the nonlinear function $\mu(\alpha) = \mu_u$ where the reduced moments in pivot $A1$, pivot $A2$ and pivot B are given by equations [3.121], [3.127] and [3.132] of [CAS 12]. The boundary between each pivot is given by:

$$\mu_{A1A2} = \frac{\alpha_{A_1A_2}^2}{2(1-\alpha_{A_1A_2})}\left(1 - \frac{\alpha_{A_1A_2}}{3}\right)\left(\frac{\varepsilon_{ud}}{-\varepsilon_{c3}}\right) = 0.01848 \quad \text{and}$$

$$\mu_{AB} = \alpha_{AB}\left(1 - \frac{\alpha_{AB}}{2}\right) - \left(\frac{\varepsilon_{c3}}{\varepsilon_{cu3}}\right)\frac{\alpha_{AB}(1-\alpha_{AB})}{2} - \frac{\alpha_{AB}^2}{6}\left(\frac{\varepsilon_{c3}}{\varepsilon_{cu3}}\right)^2 = 0.05260 \quad [1.193]$$

As $\mu_u \in [\mu_{A1A2}; \mu_{AB}]$, the behavior of the section is controlled by pivot $A2$ (see also Figure 1.19).

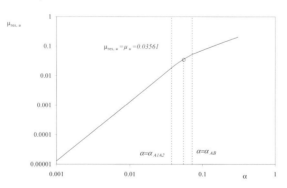

Figure 1.19. *Graphical representation of the reduced moment μ with respect to the dimensionless position of the neutral axis α*

Using equation [3.127] of [CAS 12] with $\mu_{A2}(\alpha) = \mu_u$, we obtain a second-order differential equation for α as:

$$\alpha^2 - 2\alpha + \frac{\dfrac{\varepsilon_{c3}}{2\varepsilon_{ud}} - \dfrac{1}{6}\left(\dfrac{\varepsilon_{c3}}{\varepsilon_{ud}}\right)^2 - \mu_u}{\dfrac{\varepsilon_{c3}}{2\varepsilon_{ud}} - \dfrac{1}{6}\left(\dfrac{\varepsilon_{c3}}{\varepsilon_{ud}}\right)^2 - \dfrac{1}{2}} = 0 \Rightarrow$$

$$\alpha = 1 - \sqrt{\frac{1-2\mu}{1 - \dfrac{\varepsilon_{c3}}{\varepsilon_{ud}} + \dfrac{1}{3}\left(\dfrac{\varepsilon_{c3}}{\varepsilon_{ud}}\right)^2}} = 0.05471 \qquad [1.194]$$

The solution is $\alpha = 0.05471$. It can be checked that the neutral axis is in the flanged portion of the cross-section, as $y = 0.0222$ m (largely $< h_0$). The strain value of the upper fiber of the compression concrete block can be deduced as $\varepsilon_{csup} = -\alpha\varepsilon_{ud}/(1 - \alpha) = -2.60\%_0$. The steel stress value is then directly estimated on the curve of the steel constitutive law, and is equal in pivot A to $\sigma_{s1} = q\varepsilon_{ud} + q' = 465.93$ MPa. The rebars section is calculated using equations [3.110] and [3.128] of [CAS 12] as:

$$A_{s1} = \frac{-\alpha\psi_{A2}bdf_{cu}}{\sigma_{s1}} = 12.00 \text{ cm}^2 \text{ with}$$

$$\alpha\psi_{A2} = \alpha - \frac{1-\alpha}{2}\left(\frac{-\varepsilon_{c3}}{\varepsilon_{ud}}\right) = 0.0363294 \qquad [1.195]$$

The tensile steel reinforcement area of $A_{s1} = 12.00$ cm² is needed to verify the ULS of this cross-section for the bending solicitation $M_{Ed} = 222$ kN.m. Using the abacus of Appendix 2, we can choose $4\phi20$ (12.57 cm²), which gives an upper bound of the minimum tensile steel area of 12.00 cm². However, the real value of d for this reinforced concrete section is:

$$d = h - C_{nom} - \Phi_t - \Phi_{moy} \qquad [1.196]$$

where Φ_t is the transversal rebars diameter that can be chosen as $\Phi_t = 10$ mm (assumption to verify with shear study; typically $\Phi_t \approx \Phi_{ave}/3$). Taking into account $C_{nom} = 20 + 10 = 30$ mm, and $\Phi_{ave} = 20$ mm, we calculate $d_{real} = 390$

mm. d_{real} is not superior to the value of d used for calculus ($d = 405$ mm in the calculation). Then, the calculation must be revised with a new value of d. In fact, the choice of $4\phi20$ ($A_s = 12.57$ cm²) appears to be sufficient with $d = 0.39$ m. This can be proved by using equation [3.110] of [CAS 12], and assuming pivot $A2$:

$$\alpha\psi = -A_{s1} \cdot \sigma_{s1} / (bdf_{cu}) \quad \text{with}$$
$$\sigma_{s1} = 465.93 \text{ MPa} \Rightarrow \alpha\psi = 0.0395192 \qquad [1.197]$$

Equation [3.128] of [CAS 12] is then used to obtain α:

$$\alpha\psi = \alpha + (1-\alpha)\varepsilon_{c3} / (2\varepsilon_{ud}) \Rightarrow$$
$$\alpha = (\alpha\psi - \varepsilon_{c3} / (2\varepsilon_{ud})) / (1 - \varepsilon_{c3} / (2\varepsilon_{ud})) = 0.057839 \qquad [1.198]$$

The result confirms the pivot $A2$ behavior ($\alpha_{A1A2} < \alpha < \alpha_{AB}$). Using equation [3.127] of [CAS 12], the reduced moment is calculated in pivot $A2$ as $\mu = 0.038682$. Finally, we obtain the resistant bending moment for $4\phi20$: $M_R = -\mu bd^2 f_{cu} = 223.6$ kN.m.

As $M_R > M_{Ed}$, the selected rebars section satisfy the ULS verification of the cross-section.

1.2.3.4.3. T-cross-section behavior – M_u=2,000 kN.m

In this case, M_u is higher than M_T and the section has to be calculated as a T-cross-section in pivot B. As shown in equation [1.177], the position of the neutral axis is obtained from a cubic equation when $\alpha d \in [h_0; 2h_0]$. This cubic equation is numerically given for the problem of interest with $M_u = 2,000$ kN.m and $d = 0.405$ m as:

$$0.47222222(\alpha d)^3 - 2.49750(\alpha d)^2$$
$$+ 2.229333(\alpha d) - 0.3767111 = 0 \qquad [1.199]$$

Cardano's method (see Appendix 1) is used (with the calculated canonical parameters $p=-4.602944$ m² and $q = -3.433307$ m²) leading to the solution of interest $\alpha_1 d = 0.2217642$ m. It is checked that the solution belongs to the admissible domain of variation associated with the cubic equation [1.199], that is $\alpha d \in [h_0; 2h_0]$. The strain and the stress in the tensile steel reinforcement are then calculated from:

$$\varepsilon_{s1} = \frac{\alpha - 1}{\alpha}\varepsilon_{cu3} = 12.2825\text{\textperthousand} \Rightarrow \sigma_{s1} = q\varepsilon_{s1} + q' = 442.135 \text{ MPa} \quad [1.200]$$

The tensile steel area is then computed from equation [1.179] leading to:

$$A_{s1} = \frac{\dfrac{3}{4}\alpha bd(-f_{cu}) - (b - b_w)(-f_{cu})\dfrac{(\alpha d - h_0)^2}{\alpha d}}{f_{su}} = 141.3 \text{ cm}^2 \quad [1.201]$$

The section is, in fact, not well designed for such a solicitation, as the steel ratio is close to the admissible steel ratio defined by:

$$A_{s,\max} = 0.04A_c = 0.04(b_w h + (b - b_w)h_0) = 202.4 \text{ cm}^2 \quad [1.202]$$

A modification of the geometric characteristics of the cross-section would be probably required in this case for such a solicitation.

1.2.4. T-cross-section – general equations for composed bending with normal forces

The results for the parabola–rectangle concrete law can be found [CAS 89].

1.2.4.1. Methodology

We need to decompose the stress variation into a linearly decreasing variation and a constant variation (see Figure 3.29). χd is the distance from the neutral axis to the fiber associated with the transition from the elastic part to the constant part (or plastic part) of the compression block.

According to the position of the point of transition between the linear and the constant part of the stress in the section of concrete, we can consider three cases.

The point of junction may be:

– outside of the section, the concrete is then in the elastic phase;

– in the flange of the T-cross-section;

– in the web of the T-cross-section.

Let us call χ (obtained with a division by effective depth), the relative depth of the point of transition from the neutral axis. This variable can take three positives values depending on the pivot: χ_{A1}, χ_{A2} and χ_B:

$$\chi_{A1} = \frac{-(1-\alpha).\varepsilon_{c3}}{\varepsilon_{ud}}; \; \chi_{A2} = \frac{-(1-\alpha).\varepsilon_{c3}}{\varepsilon_{ud}}; \; \chi_B = \frac{\varepsilon_{c3}.\alpha}{\varepsilon_{cu3}} \qquad [1.203]$$

with the following inequalities:

$$0 \le \alpha \le \alpha_{A1A2} = \frac{\varepsilon_{c3}}{\varepsilon_{c3}-\varepsilon_{ud}} \text{ and: } \frac{-\varepsilon_{c3}}{\varepsilon_{ud}} \ge \chi_{A1} \ge \alpha_{A1A2} = \frac{\varepsilon_{c3}}{\varepsilon_{c3}-\varepsilon_{ud}}$$

$$\alpha_{A1A2} = \frac{\varepsilon_{c3}}{\varepsilon_{c3}-\varepsilon_{ud}} \le \alpha \le \alpha_{AB} = \frac{\varepsilon_{cu3}}{\varepsilon_{cu3}-\varepsilon_{ud}} \text{ and:}$$

$$\alpha_{A1A2} = \frac{\varepsilon_{c3}}{\varepsilon_{c3}-\varepsilon_{ud}} \ge \chi_{A2} \ge \frac{\alpha_{AB}.\varepsilon_{c3}}{\varepsilon_{cu3}} = \frac{\varepsilon_{c3}}{\varepsilon_{cu3}-\varepsilon_{ud}}$$

$$\alpha_{AB} = \frac{\varepsilon_{cu3}}{\varepsilon_{cu3}-\varepsilon_{ud}} \le \alpha \le \frac{h}{d} \text{ and: } \frac{\alpha_{AB}.\varepsilon_{c3}}{\varepsilon_{cu3}} = \frac{\varepsilon_{c3}}{\varepsilon_{cu3}-\varepsilon_{ud}} \le \chi_B \le \frac{\varepsilon_{c3}.h}{\varepsilon_{cu3}.d} \qquad [1.204]$$

Let us remember the values of m and m' that have been defined in equations [3.89]-[3.91] of [CAS 12]:

$$\left.\begin{array}{l} m = q \text{ and } m'=q' \text{ for } \varepsilon_s \in [\varepsilon_{su}; \varepsilon_{ud}] \\ m = E_s \text{ and } m'=0 \text{ for } \varepsilon_s \in [-\varepsilon_{su}; \varepsilon_{su}] \\ m = q \text{ and } m'=-q' \text{ for } \varepsilon_s \in [-\varepsilon_{ud}; -\varepsilon_{su}] \end{array}\right\} \qquad [1.205]$$

and the plasticity hardening parameters q and q' are:

$$q = \frac{(k-1)f_{su}}{\varepsilon_{uk} - \varepsilon_{su}}; \; q' = f_{su}\left(1 - \frac{q}{E_s}\right) \text{ and } \varepsilon_{su} = \frac{f_{su}}{E_s} \qquad [1.206]$$

We have for the calculation of the stress in the steel reinforcements:

in pivot A: $\sigma_{s1} = k.f_{su}$ and $\sigma_{s2} = -m_{s2}.\varepsilon_{ud}\dfrac{\alpha-\delta'}{1-\alpha} + m'_{s2}$ \qquad [1.207]

$$\text{in pivot } B: \sigma_{s1} = - m_{s1}.\varepsilon_{cu3} \frac{1-\alpha}{\alpha} + m'_{s1}$$

$$\text{and } \sigma_{s2} = m_{s2}.\varepsilon_{cu3} \frac{\alpha-\delta'}{\alpha} + m'_{s2} \qquad [1.208]$$

We will define η_0 as the geometric ratio h_0/d where h_0 is the T-section flange's thickness. As for the other parts of this book, we use some dimensionless parameters as:

$$\mu_u = \frac{-M_{u,act}}{b.d^2.f_{cu}}; \ \nu_u = \frac{-N_{u,act}}{b.d.f_{cu}}; \ \rho'_{si} = \frac{A_{si}}{bd};$$

$$m'_{psi} = \frac{-m'_{si}}{f_{cu}} \text{ and } m_{psi} = \frac{-m_{si}}{f_{cu}} \qquad [1.209]$$

The determination of the parameters depends on the position of the junction of the two parts of the bilinear law, with respect to the flange junction:

– First, in pivot A_1, the strain of the upper fiber of concrete is greater than ε_{c3} (algebraically negative), the concrete is then in the elastic phase (see Figure 1.20(a)). In this case, the neutral axis may be located above the flange or below.

– On the other hand, for pivots A_2 and B, two cases must be distinguished:

– $\chi \geq \alpha - \eta_0$ the junction of the two parts of the bilinear law is in the flange (see Figure 1.20(b)).

– $\chi < \alpha - \eta_0$ the junction of the two parts of the bilinear law is in the rib (see Figure 1.20(c)).

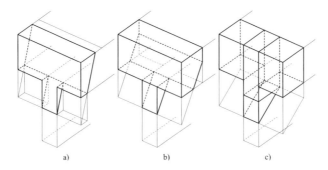

a)   b)   c)

Figure 1.20. *The three positions of the crossover point.*

1.2.4.2. *Pivot A1, junction's point outside of T-cross-section*

As shown in Figure 1.20(a), for the case $\chi \geq \alpha$ and $\alpha \geq \eta_0$, we decompose the concrete resultant measured at the center of gravity of the most tensioned reinforcements into four parts:

– The first part is created by a linear distribution of the stress applied to a rectangular section with a width equal to the wide flange b and a depth at least equal to αd; this torque will be taken as positive.

– The second part is created by a linear distribution of the stress applied from the flange's bottom to the neutral axis to a rectangular section of a width equal to $b - b_w$; this torque will be taken off.

– The third part is created by the compressed reinforcements.

– The fourth part is created by the tensile reinforcement.

In what follows, the moments are taken from the center of gravity of the tensile reinforcement.

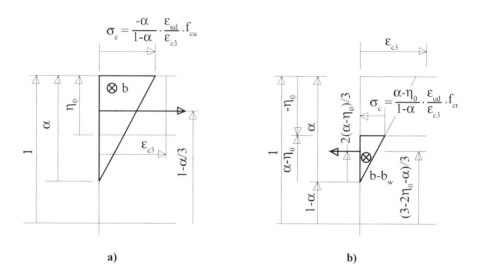

Figure 1.21.

In Figure 1.21(a), we can calculate the strain in the upper fiber as:

$$\frac{-\alpha}{1-\alpha} \cdot \frac{\varepsilon_{ud}}{\varepsilon_{c3}} \cdot f_{cu} \qquad [1.210]$$

The internal resultant force is created by a linear distribution of stress applied over a rectangular section with a width b equal to the width of the flange. The inner force is evaluated to:

$$\frac{-\alpha^2 .b.d}{2(1-\alpha)} \cdot \frac{\varepsilon_{ud} \cdot f_{cu}}{\varepsilon_{c3}} \qquad [1.211]$$

its lever arm is:

$$(1-\alpha/3) \cdot d \qquad [1.212]$$

and the inner moment is:

$$\frac{\alpha^2 .(3-\alpha)}{6(1-\alpha)} \cdot \frac{\varepsilon_{ud}}{\varepsilon_{c3}} .b.d^2 .f_{cu} \qquad [1.213]$$

In Figure 1.21(b), we can calculate the upper strain of the linear distribution (at the bottom of the flange) as:

$$\frac{\alpha-\eta_0}{1-\alpha} \cdot \frac{\varepsilon_{ud}}{\varepsilon_{c3}} \cdot f_{cu} \qquad [1.214]$$

The internal resultant force created by the linear distribution of stress applied from the flange's bottom to the neutral axis to a rectangular section of width equal to $b-b_w$ is:

$$\frac{(\alpha-\eta_0)^2 .(1-\beta).b.d}{2(1-\alpha)} \cdot \frac{\varepsilon_{ud} \cdot f_{cu}}{\varepsilon_{c3}} \qquad [1.215]$$

its lever arm is:

$$(3-2\cdot\eta_0 -\alpha) \cdot d/3 \qquad [1.216]$$

and the inner moment is:

$$-\frac{(\alpha-\eta_0)^2 .(1-\beta).(3-2.\eta_0 -\alpha).b.d^2}{6(1-\alpha)} \cdot \frac{\varepsilon_{ud} \cdot f_{cu}}{\varepsilon_{c3}} \qquad [1.217]$$

The internal moment taken by compression reinforcements is:

$$- A_{s2} \cdot \left(- m_{s2} \cdot \varepsilon_{ud} \frac{\alpha - \delta'}{1 - \alpha} + m'_{s2} \right)(d - d') \qquad [1.218]$$

Finally, the moment induced by the internal tension in the tensile steel reinforcements is zero. The balance between the acting and resisting moments leads to the following equation:

$$\alpha^3 . \beta . \varepsilon_{ud} - 3 . \alpha^2 . \beta . \varepsilon_{ud}$$
$$- \alpha . \{ 3 . (1 - \beta) . \varepsilon_{ud} . (2 - \eta_0) . \eta_0 + 6 . \varepsilon_{c3} . [\mu_u + (1 - \delta') . (m'_{ps2} + m_{ps2} . \varepsilon_{s2}) . \rho'_{s2}] \} + \dots$$
$$\dots + (1 - \beta) . \varepsilon_{ud} . \eta_0^2 . (3 - 2 . \eta_0) - 6 . \varepsilon_{c3} . \mu_u - 6 . (1 - \delta') . \varepsilon_{c3} . (m'_{ps2} + m_{ps2} . \varepsilon_{s2}) . \rho'_{s2} = 0$$

$$[1.219]$$

This equation is a polynomial of degree 3 in α. The degree of this equation is unchanged even in the case of a rectangular section ($\beta = 1$, $\eta_0 = 0$) without compressed steel reinforcement ($\rho_{s2} = 0$).

The coefficients of this polynomial of α are:

– Coefficient of degree 3

$$\beta \cdot \varepsilon_{ud} \qquad [1.220]$$

– Coefficient of degree 2

$$-3 \cdot \beta \cdot \varepsilon_{ud} \qquad [1.221]$$

– Coefficient of degree 1

$$-3 . (1 - \beta) . \varepsilon_{ud} . (2 - \eta_0) . \eta_0 + 6 . \varepsilon_{c3} . [\mu_u + (1 - \delta') . (m'_{ps2} + m_{ps2} . \varepsilon_{s2}) . \rho'_{s2}] . \qquad [1.222]$$

– Coefficient of degree 0

$$(1 - \beta) . \varepsilon_{ud} . \eta_0^2 . (3 - 2 . \eta_0) - 6 . \varepsilon_{c3} . \mu_u - 6 . (1 - \delta') . \varepsilon_{c3} . (m'_{ps2} + m_{ps2} . \varepsilon_{s2}) . \rho'_{s2} \qquad [1.223]$$

The balance between the acting and resisting axial forces leads to the following equation:

$$-V_u + \frac{\alpha^2.\varepsilon_{ud}-(1-\beta).\varepsilon_{ud}.(\alpha-\eta_0)^2}{2.(1-\alpha).\varepsilon_{c3}} + (m'_{ps1}+m_{ps1}.\varepsilon_{s1}).\rho'_{s1} +$$

$$(m'_{ps2}+m_{ps2}.\varepsilon_{s2}).\rho'_{s2} = 0 \qquad\qquad [1.224]$$

This can then determine ρ'_{s1}:

$$\rho'_{s1} = \frac{2.V_u.(1-\alpha).\varepsilon_{c3}-\varepsilon_{ud}.[\alpha^2.\beta+(1-\beta).\eta_0.(2.\alpha-\eta_0)]-2.(1-\alpha).\varepsilon_{c3}.(m'_{ps2}+m_{ps2}.\varepsilon_{s2}).\rho'_{s2}}{2.(1-\alpha).\varepsilon_{c3}.(m'_{ps1}+m_{ps1}.\varepsilon_{s1})}$$

$$[1.225]$$

1.2.4.3. *Junction's point inside T-cross-section's flange*

1.2.4.3.1. Introduction

As shown in Figure 1.20(b), for the case $\chi \geq \alpha - \eta_0$, we decompose the torque forces measured at the center of gravity of the most tensioned reinforcements into five parts:

–The first part is created by a rectangular distribution of the stress applied over the entire flange; this torque will be taken as positive.

–The second part is created by a linear distribution of the stress applied from the junction between the two parts of the bilinear law to the flange's bottom; this torque will be taken off.

–The third part is created by a linear distribution applied to the web of T-cross-section from the bottom of the flange to the neutral axes.

–The fourth part is created by the compressed reinforcements.

–The fifth part is created by the tensile reinforcement.

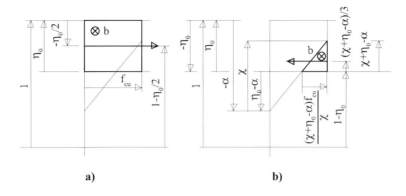

Figure 1.22.

In Figure 1.22(a), in which all vertical dimensions have been divided by *d*, we can calculate the internal resultant force created by the rectangular distribution of stress applied over the full height of the flange. The inner force is evaluated to:

$$\eta_0 \cdot b \cdot d \cdot f_{cu} \qquad [1.226]$$

Its lever arm is:

$$(1 - \eta_0 / 2) \cdot d \qquad [1.227]$$

The inner moment is:

$$-\eta_0 \cdot (1 - \eta_0 / 2) \cdot b \cdot d^2 \cdot f_{cu} \qquad [1.228]$$

With Figure 1.22(b), the internal resultant force created by the linear distribution of the stress applied from the junction between the two parts of the bilinear law to the flange's bottom is:

$$-\frac{(\chi + \eta_0 - \alpha)^2 . b . d . f_{cu}}{2 . \chi} \qquad [1.229]$$

Its lever arm is:

$$(3 - \alpha - 2\eta_0 + \chi) \cdot d / 3 \qquad [1.230]$$

The inner moment is:

$$\frac{(\chi + \eta_0 - \alpha)^2 (3 - \alpha - 2\eta_0 + \chi).b.d^2.f_{cu}}{6.\chi}$$

$$[1.231]$$

The internal moment taken by the compression reinforcements is:

$$- A_{s2}.(m_{s2}.\varepsilon_{s2} + m'_{s2})(d - d')$$

$$[1.232]$$

The internal moment for the steel tensile reinforcements is zero. The balance between the acting and resisting moments leads to the following equation:

$$(1 - \beta).(\alpha - \eta_0)^2.(-3 + \alpha + 2.\eta_0)$$
$$+ 3.[(2 - \alpha).\alpha - 2.\mu_u + 2.(1 - \delta').(m'_{ps2} + m_{ps2}.\varepsilon_{s2}).\rho'_{s2}].\chi$$
$$- 3.(1 - \alpha).\chi^2 - \chi^3 = 0$$

$$[1.233]$$

The balance between the acting and resisting axial forces leads to the following equation:

$$- v_u - \eta_0 + (m'_{ps1} + m_{ps1}.\varepsilon_{s1}).\rho'_{s1} + (m'_{ps2} + m_{ps2}.\varepsilon_{s2}).\rho'_{s2}$$

$$+ \frac{(\alpha - \eta_0 - \chi)^2 - \beta.(\alpha - \eta_0)^2}{2.\chi} = 0$$

$$[1.234]$$

Now, we have only to substitute χ by its expressions in pivot $A2$ or B.

1.2.4.3.2. Pivot $A2$, junction's point inside T-cross-section's flange

The balance between the acting and resisting moments leads to the following equation:

$$(1 - \alpha)^3.\varepsilon_{c3}^2(\varepsilon_{c3} - 3.\varepsilon_{ud}) - (1 - \beta).\varepsilon_{ud}^3.(\alpha - \eta_0)^2.(3 - \alpha - 2.\eta_0)$$

$$- 3.\varepsilon_{c3}.\varepsilon_{ud}.\{(1 - \alpha).[(2 - \alpha).\alpha - 2.\mu_u] - 2.(1 - \delta').[m'_{ps2} - m_{ps2}.\alpha - m_{ps2}.(\alpha - \delta').\varepsilon_{ud}].\rho'_{s2}\} = 0$$

$$[1.235]$$

This equation is a polynomial of degree 3 in α; the coefficients of this polynomial of α are:

– Coefficient of degree 3

$$-\varepsilon_{c3}^2(\varepsilon_{c3} - 3.\varepsilon_{ud}) + \varepsilon_{ud}^2[(1 - \beta).\varepsilon_{ud} - 3.\varepsilon_{c3}]$$

$$[1.236]$$

– Coefficient of degree 2

$$3.\{\varepsilon^2_{c3}(\varepsilon_{c3}-3.\varepsilon_{ud})-\varepsilon^2_{ud}[(1-\beta).\varepsilon_{ud}-3.\varepsilon_{c3}]\} \qquad [1.237]$$

– Coefficient of degree 1

$$-3.\varepsilon^2_{c3}(\varepsilon_{c3}-3.\varepsilon_{ud})+3.(1-\beta).\varepsilon^3_{ud}.(2-\eta_0).\eta_0-6.\varepsilon_{c3}.\varepsilon^2_{ud}.$$
$$[1+\mu_u+(1-\delta')(m'_{ps2}+m_{ps2}.\varepsilon_{ud}).\rho'_{s2}] \qquad [1.238]$$

– Coefficient of degree 0

$$\varepsilon^2_{c3}(\varepsilon_{c3}-3.\varepsilon_{ud})-(1-\beta).\varepsilon^3_{ud}.\eta_0^2.(3-2.\eta_0)+6.\varepsilon_{c3}.\varepsilon^2_{ud}.$$
$$[\mu_u+(1-\delta').(m'_{ps2}+m_{ps2}.\delta'.\varepsilon_{ud}).\rho'_{s2}] \qquad [1.239]$$

The degree of this equation is unchanged even in the case of a rectangular section ($\beta = 1$, $\eta_0 = 0$) without compressed reinforcement ($\rho'_{s2} = 0$). The balance between the acting and resisting axial forces leads to the following equation:

$$-\nu_u+\varepsilon_{ud}.\frac{\beta.(\alpha-\eta_0)^2-\left[\alpha+\dfrac{(1-\alpha).\varepsilon_{c3}}{\varepsilon_{ud}}-\eta_0\right]^2}{2.(1-\alpha).\varepsilon_{c3}}-\eta_0+(m'_{ps1}+m_{ps1}.\varepsilon_{ud}).\rho'_{s1}$$
$$+\left[m'_{ps2}-\frac{m_{ps2}.(\alpha-\delta').\varepsilon_{ud}}{1-\alpha}\right].\rho'_{s2}=0 \qquad [1.240]$$

This can then determine ρ'_{s1}:

$$\rho'_{s1}=\frac{(1-\alpha)^2\varepsilon^2_{c3}+(1-\beta)\varepsilon^2_{ud}(\alpha-\eta_0)^2+2\varepsilon_{c3}\varepsilon_{ud}\{(1-\alpha)(\nu_u+\alpha)-[m'_{ps2}(1-\alpha)-m_{ps2}(\alpha-\delta')\varepsilon_{ud}]\rho'_{s2}}{2(1-\alpha)\varepsilon_{c3}\varepsilon_{ud}(m'_{ps1}+m_{ps1}\varepsilon_{ud})}$$

$$[1.241]$$

1.2.4.3.3. Pivot B, junction's point inside T-cross-section's flange

The balance between the acting and resisting moments leads to the following equation:

$$\alpha^3.[-\varepsilon^3_{c3}+3.\varepsilon^2_{c3}.\varepsilon_{cu3}-3.\varepsilon_{c3}.\varepsilon^2_{cu3}+(1-\beta).\varepsilon^3_{cu3}]-3.\alpha^2.\varepsilon_{cu3}.$$
$$[\varepsilon^2_{c3}-2.\varepsilon_{c3}.\varepsilon_{cu3}+(1-\beta).\varepsilon^2_{cu3}]+3.\alpha.\varepsilon^2_{cu3}.\{(1-\beta).\varepsilon_{cu3}.(2-\eta_0).$$
$$\eta_0-2.m_{ps2}.(1-\delta').\varepsilon_{c3}.\varepsilon_{cu3}.\rho_{s2}-2.\varepsilon_{c3}.[\mu_u+m'_{ps2}.(1-\delta').\rho'_{s2}]\}-\varepsilon^3_{cu3}.$$
$$[(1-\beta).\eta_0^2.(3-2.\eta_0)-6.m_{ps2}.(1-\delta').\delta'.\varepsilon_{c3}.\rho'_{s2}] = 0 \qquad [1.242]$$

This equation is exactly equation [1.176] with $\varepsilon_{cu3} = 2.\varepsilon_{c3}$. This equation is a polynomial of degree 3 in α, the coefficients of this polynomial of α are:

– Coefficient of degree 3

$$-\varepsilon'^3_{c3}+3.\varepsilon'^2_{c3}.\varepsilon_{cu3}-3.\varepsilon_{c3}.\varepsilon'^2_{cu3}+(1-\beta).\varepsilon'^3_{cu3} \qquad [1.243]$$

– Coefficient of degree 2

$$-3.\varepsilon_{cu3}.[\varepsilon'^2_{c3}-2.\varepsilon_{c3}.\varepsilon_{cu3}+(1-\beta).\varepsilon'^2_{cu3}] \qquad [1.244]$$

– Coefficient of degree 1

$$3.\varepsilon'^2_{cu3}.\{(1-\beta).\varepsilon_{cu3}.(2-\eta_0).\eta_0-2.m_{ps2}.(1-\delta').\varepsilon_{c3}.\varepsilon_{cu3}.$$
$$\rho'_{s2}-2.\varepsilon_{c3}.[\mu_u+m'_{ps2}.(1-\delta').\rho'_{s2}]\} \qquad [1.245]$$

– Coefficient of degree 0

$$-\varepsilon^3_{cu3}.[(1-\beta).\eta_0^2.(3-2.\eta_0)-6.m_{ps2}.(1-\delta').\delta'.\varepsilon_{c3}.\rho'_{s2}] \qquad [1.246]$$

The degree of this equation is changed in the case of a rectangular section ($\beta = 1$, $\eta_0 = 0$) without compressed reinforcement ($\rho'_{s2} = 0$). The polynomial is of degree 2 in this case.

– Coefficient of degree 2

$$-\varepsilon'^3_{c3}+3.\varepsilon'^2_{c3}.\varepsilon_{cu3}-3.\varepsilon_{c3}.\varepsilon'^2_{cu3} \qquad [1.247]$$

– Coefficient of degree 1

$$-3.\varepsilon_{cu3}.(\varepsilon'^2_{c3}-2.\varepsilon_{c3}.\varepsilon_{cu3}) \qquad [1.248]$$

– Coefficient of degree 0

$$-6.\varepsilon_{cu3}^2.\varepsilon_{c3}.\mu_u \qquad [1.249]$$

The balance between the acting and resisting axial forces leads to the following equation:

$$\nu_u+\varepsilon_{cu3}.\frac{\beta.(\alpha-\eta_0)^2.\left[\alpha-\dfrac{\alpha.\varepsilon_{c3}}{\varepsilon_{cu3}}-\eta_0\right]^2}{2.\alpha.\varepsilon_{c3}}+\eta_0-\left[m'_{ps1}-\frac{m_{ps1}.(1-\alpha).\varepsilon_{cu3}}{\alpha}\right].\rho'_{s1}$$
$$-\left[m'_{ps2}+\frac{m_{ps2}.(\alpha-\delta').\varepsilon_{cu3}}{\alpha}\right].\rho'_{s2}=0 \qquad [1.250]$$

This can then determine ρ'_{s1} :

$$\rho'_{s1} = \frac{-\alpha^2.[\varepsilon^2_{c3}-2.\varepsilon_{c3}.\varepsilon_{cu3}+(1-\beta).\varepsilon^2_{cu3}]-\varepsilon^2_{cu3}.[(1-\beta).\eta_0^2-2.m_{ps2}.\delta'.\varepsilon_{c3}.\rho'_{s2}]}{2.\varepsilon_{c3}.\varepsilon_{cu3}.[m'_{ps1}.\alpha-m_{ps1}.(1-\alpha).\varepsilon_{cu3}]} +$$

$$+\frac{2.\alpha.\varepsilon_{cu3}.[\nu_u.\varepsilon_{c3}-(-1+\beta).\varepsilon_{cu3}.\eta_0-\varepsilon_{c3}.(m'_{ps2}+m_{ps2}.\varepsilon_{cu3}).\rho'_{s2}]}{2.\varepsilon_{c3}.\varepsilon_{cu3}.[m'_{ps1}.\alpha-m_{ps1}.(1-\alpha).\varepsilon_{cu3}]} \qquad [1.251]$$

In the case of a rectangular section ($\beta = 1$, $\eta_0 = 0$) without compressed reinforcement ($\rho'_{s2} = 0$), we can found:

$$\rho'_{s1} = \frac{-\alpha^2.(\varepsilon_{c3}-2.\varepsilon_{cu3})+2.\alpha.\varepsilon_{cu3}.\nu_u}{2.\varepsilon_{cu3}.[m'_{ps1}.\alpha-m_{ps1}.(1-\alpha).\varepsilon_{cu3}]} \qquad [1.252]$$

1.2.4.4. *Junction's point inside T-cross-section's web*

1.2.4.4.1. Introduction

As shown in Figure 1.20(c), for the case $\chi \geq \alpha - \eta_0$, we decompose the torque forces measured at the center of gravity of the most tensioned reinforcements into five parts:

–The first part is created by a rectangular distribution of the stress applied over the two sides of the flange; this torque will be taken as positive.

–The second part is created by a rectangular distribution of the stress applied from the top of the web to the junction between the two parts of the bilinear law; this torque will be taken off.

–The third part is created by a linear distribution applied to the web of T-cross-section from the junction point to the neutral axes.

–The fourth part is created by the compressed reinforcements.

–The fifth part is created by the tensile reinforcement.

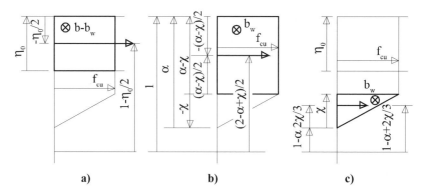

Figure 1.23.

With Figure 1.23(a), the internal resultant force created by the rectangular distribution of stress applied to the web of T-section from the top to the bottom of the flange is:

$$\eta_0 \cdot (1-\beta) \cdot b \cdot d \cdot f_{cu} \qquad [1.253]$$

Its lever arm is:

$$(2-\eta_0) \cdot d / 2 \qquad [1.254]$$

The inner moment is:

$$-\eta_0.(1-\beta).(1-\eta_0/2).b.d^2.f_{cu} \qquad [1.255]$$

With Figure 1.23(b), the internal resultant force created by the rectangular distribution of stress applied to the web of T-section from the top of this section to the junction point is:

$$(\alpha-\chi) \cdot b_w \cdot d \cdot f_{cu} \qquad [1.256]$$

Its lever arm is:

$$(2-\alpha+\chi) \cdot d / 2 \qquad [1.257]$$

The inner moment is:

$$-(\alpha-\chi) \cdot (2-\alpha+\chi) \cdot \beta \cdot b \cdot d^2 \cdot f_{cu} \qquad [1.258]$$

With Figure 1.23 (c), the internal resultant force created by the linear distribution of stress applied to the web of T-section from the junction point to the neutral axes is:

$$\chi \cdot b_w \cdot d \cdot f_{cu} \tag{1.259}$$

Its lever arm is:

$$(1 - \alpha + 2 \cdot \chi) \cdot d / 3 \tag{1.260}$$

The inner moment is:

$$-\chi \cdot (1 - \alpha + 2 \cdot \chi) \cdot \beta \cdot b \cdot d^2 \cdot f_{cu} \tag{1.261}$$

The internal moment taken by compression reinforcements is:

$$- A_{s2}.(m_{s2}.\varepsilon_{s2} + m'_{s2})(d{-}d') \tag{1.262}$$

The internal moment taken by tension reinforcements is zero.

The resulting equilibrium relationship in moment is:

$$3.\alpha^2.\beta{-}3.(1{-}\beta).(2{-}\eta_0).\eta_0{+}6.\mu_u{+}6.(1{-}\delta').(m'_{ps2}{+}m_{ps2}.\varepsilon_{s2}).\rho'_{s2}$$
$$+3.\beta.\chi{+}\beta.\chi^2{-}3.\alpha.\beta.(2{+}\chi) = 0 \tag{1.263}$$

The balance between the acting and resisting axial forces leads to the following equation:

$$v_u{+}\alpha.\beta{+}(1{-}\beta).\eta_0{-}m'_{ps1}.\rho'_{s1}{-}m_{ps1}.\varepsilon_{s1}.\rho'_{s1}$$
$$-m'_{ps2}.\rho'_{s2}{-}m_{ps2}.\varepsilon_{s2}.\rho'_{s2}{-}(\beta.\chi)/2 = 0 \tag{1.264}$$

Now, we have only to substitute χ by its expressions in pivot $A2$ or B.

1.2.4.4.2. Pivot $A2$, junction's point inside T-cross-section's web

The balance between the acting and resisting moments leads to the following equation:

$$(1{-}\alpha).\{\beta.[(1{-}\alpha)^2.\varepsilon^2_{c3}{-}3.(1{-}\alpha)^2.\varepsilon_{c3}.\varepsilon_{ud}{-}3.\varepsilon^2_{ud}.(\alpha{-}\eta_0).$$
$$(2{-}\alpha{-}\eta_0)]{-}3.\varepsilon^2_{ud}.[(2{-}\eta_0).\eta_0{-}2.\mu_u]\}{+}6.(1{-}\delta').\varepsilon^2_{ud}.$$
$$[m'_{ps2}{-}m'_{ps2}.\alpha{-}m_{ps2}.(\alpha{-}\delta').\varepsilon_{ud}].\rho'_{s2} = 0 \tag{1.265}$$

It is a polynomial of degree 3 whose coefficients are:

– Coefficient of degree 3

$$\beta.(\varepsilon^2_{c3}-3.\varepsilon_{c3}.\varepsilon_{ud}+3.\varepsilon^2_{ud}) \qquad\qquad [1.266]$$

– Coefficient of degree 2

$$-3.\beta.(\varepsilon^2_{c3}-3.\varepsilon_{c3}.\varepsilon_{ud}+3.\varepsilon^2_{ud}) \qquad\qquad [1.267]$$

– Coefficient of degree 1

$$3.\beta.[\varepsilon^2_{c3}-3.\varepsilon_{c3}.\varepsilon_{ud}+\varepsilon^2_{ud}.(2+2.\eta_0-\eta_0^2)]+3.\varepsilon^2_{ud}.[2.\eta_0-\eta_0^2-2.$$
$$(\mu_u+m'_{ps2}.\rho'_{s2}-m'_{ps2}.\delta'.\rho'_{s2}+m_{ps2}.\varepsilon_{ud}.\rho'_{s2}-m_{ps2}.\delta'.\varepsilon_{ud}.\rho'_{s2})] \qquad [1.268]$$

– Coefficient of degree 0

$$-\beta.[\varepsilon^2_{c3}-3.\varepsilon_{c3}.\varepsilon_{ud}+3.\varepsilon^2_{ud}.(2-\eta_0).\eta_0]+3.\varepsilon^2_{ud}.$$
$$[-(2-\eta_0).\eta_0+2.\mu_u+2.(1-\delta').(m'_{ps2}+m_{ps2}.\delta'.\varepsilon_{ud}).\rho'_{s2}] \qquad [1.269]$$

The degree of this equation is unchanged in the case of a rectangular section ($\beta = 1$, $\eta_0 = 0$) without compressed reinforcement ($\rho'_{s2} = 0$). The balance between the acting and resisting axial forces leads to the following equation:

$$\nu_u+\alpha.\beta+\frac{(1-\alpha).\beta.\varepsilon_{c3}}{2.\varepsilon_{ud}}+(1-\beta).\eta_0-m'_{ps1}.\rho'_{s1}-m_{ps1}.\varepsilon_{ud}.\rho'_{s1}$$
$$-m'_{ps2}.\rho'_{s2}+\frac{m_{ps2}.(\alpha-\delta').\varepsilon_{ud}.\rho'_{s2}}{1-\alpha}=0 \qquad [1.270]$$

which can give the tensile reinforcement:

$$\rho'_{s1}=\frac{\nu_u+\beta+\dfrac{(1-\alpha).\beta.\varepsilon_{c3}}{2.\varepsilon_{ud}}+\eta_0.(1-\beta)-m'_{ps2}.\rho'_{s2}+\dfrac{m_{ps2}.(\alpha-\delta').\varepsilon_{ud}.\rho'_{s2}}{1-\alpha}}{m'_{ps1}+m_{ps1}.\varepsilon_{ud}} \qquad [1.271]$$

1.2.4.4.3. Pivot B, junction's point inside T-cross-section's web

In this case, the balance between the acting and resisting moments leads to the following equation:

$$\alpha^3.\beta.(\varepsilon^2_{c3}-3.\varepsilon_{c3}.\varepsilon_{cu3}+3.\varepsilon^2_{cu3})+3.\alpha^2.\beta.(\varepsilon_{c3}-2.\varepsilon_{cu3}).\varepsilon_{cu3}-3.\alpha.\varepsilon^2_{cu3}.[(1-$$
$$\beta).(2-\eta_0).\eta_0-2.\mu_u-2.m'_{ps2}.(1-\delta').\rho'_{s2}-2.m_{ps2}.(1-\delta').\varepsilon_{cu3}.\rho'_{s2}]-$$
$$6.m_{ps2}.(1-\delta').\delta'.\varepsilon^3_{cu3}.\rho'_{s2} = 0 \qquad [1.272]$$

There is also a polynomial of degree 3 whose coefficients are:

– Coefficient of degree 3

$$\beta.(\varepsilon^2_{c3}-3.\varepsilon_{c3}.\varepsilon_{cu3}+3.\varepsilon^2_{cu3}) \qquad [1.273]$$

– Coefficient of degree 2

$$3.\beta.(\varepsilon_{c3}-2.\varepsilon_{cu3}).\varepsilon_{cu3} \qquad [1.274]$$

– Coefficient of degree 1

$$-3.\alpha.\varepsilon^2_{cu3}.[(1-\beta).(2-\eta_0).\eta_0-2.\mu_u-2.m'_{ps2}.$$
$$(1-\delta').\rho'_{s2}-2.m_{ps2}.(1-\delta').\varepsilon_{cu3}.\rho'_{s2}] \qquad [1.275]$$

– Coefficient of degree 0

$$-6.m_{ps2}.(1-\delta').\delta'.\varepsilon^3_{cu3}.\rho'_{s2} \qquad [1.276]$$

The degree of this equation is changed in the case of a rectangular section ($\beta = 1$, $\eta_0 = 0$) without compressed reinforcement ($\rho'_{s2} = 0$). The polynomial is of degree 2.

– Coefficient of degree 2

$$\varepsilon^2_{c3}-3.\varepsilon_{c3}.\varepsilon_{cu3}+3.\varepsilon^2_{cu3} \qquad [1.277]$$

– Coefficient of degree 1

$$3.(\varepsilon_{c3}-2.\varepsilon_{cu3}).\varepsilon_{cu3} \qquad [1.278]$$

– Coefficient of degree 0

$$6.\alpha.\varepsilon^2_{cu3}.\mu_u \qquad [1.279]$$

The balance between the acting and resisting axial forces leads to the following equation:

$$v_u + (1-\beta).\eta_0 - m'_{ps1}.\rho'_{s1} - m'_{ps2}.\rho'_{s2} + \frac{\alpha.\beta.(2.\varepsilon_{cu3} - \varepsilon_{c3})}{2.\varepsilon_{cu3}} +$$

$$+ \frac{m_{ps1}.(1-\alpha).\varepsilon_{cu3}.\rho'_{s1} - m_{ps2}.(\alpha-\delta').\varepsilon_{cu3}.\rho'_{s2}}{\alpha} = 0 \qquad [1.280]$$

Then, the tensile reinforcement is:

$$\rho'_{s1} = \frac{\alpha^2\beta(2\varepsilon_{cu3} - \varepsilon_{c3}) + 2\alpha\varepsilon_{cu3}[v_u + \eta_0 - \beta\eta_0 - (m'_{ps2} + m_{ps2}\varepsilon_{cu3})\rho'_{s2}] + 2m_{ps2}\delta'\varepsilon^2_{cu3}\rho'_{s2}}{2\varepsilon_{cu3}[m'_{ps1}\alpha - m_{ps1}(1-\alpha)\varepsilon_{cu3}]} \qquad [1.281]$$

1.3. ULS – interaction diagram

1.3.1. *Theoretical formulation of the interaction diagram*

1.3.1.1. *Phenomenological analysis of the interaction diagram*

For each reinforced concrete cross-section, the ultimate bending moment M_u and normal forces N_u can be calculated from the ultimate constitutive law of the steel reinforcement and concrete in compression. These internal forces can be expressed, for instance, with respect to the center of gravity of the lower steel reinforcement (the tensile steel reinforcement in case of pure bending with positive bending moment). We then obtain a parameterized curve where the dimensionless position of the neutral axis α plays the role of the varying parameter as:

$$\begin{cases} M_u(\alpha) = M_c(\alpha) - A_{s2}\sigma_{s2}(\alpha)(d-d') \\ N_u(\alpha) = N_c(\alpha) + \sigma_{s1}(\alpha)A_{s1} + \sigma_{s2}(\alpha)A_{s2} \end{cases} \quad \text{with } y = \alpha d \qquad [1.282]$$

where M_c and N_c are the internal bending moments and axial forces in the compression concrete block, calculated with respect to the center of gravity of the lower steel reinforcement. The ultimate capacity of the cross-section can also be characterized with respect to the center of gravity of the concrete section, with the rule of frame changing as:

$$\begin{cases} M_u^*(\alpha) = M_u(\alpha) + \left(d - \frac{h}{2}\right)N_u(\alpha) \\ N_u^*(\alpha) = N_u(\alpha) \end{cases} \qquad [1.283]$$

Figure 1.24 shows a typical interaction diagram for a rectangular symmetrically reinforced concrete section. The relative magnitude of the moment M and the axial load N governs whether the section will fail mostly in tension (pivot A or at least yielding of the tension steel reinforcement) with some possible strain limitation in the steel reinforcement, or in compression (pivot B) with a strain limitation in the concrete compression block (and the possible elasticity behavior of the tensile steel reinforcement). With large effective eccentricity ($e = M/N$), a tensile failure is likely with yielding of the tension steel reinforcement, but with a small eccentricity, a compression failure is more likely in pivot B, with the possible elastic behavior of the tensile steel reinforcement. The magnitude of the eccentricity affects the position of the neutral axis (as for the SLS), and hence, the strains and the stresses in the steel and compression steel reinforcements.

The change of behavior between the tension-type failure (yielding of the tensile steel reinforcement) and the compression-type failure (associated with an elastic behavior of the tensile steel reinforcement), as classified, for instance, by Paulay and Park [PAU 75] or MacGregor [MAC 97], is obtained at the balanced failure point. As detailed by Paulay and Park [PAU 75], the term "tension-type failure" expresses the fact that yielding of the tension steel reinforcement precedes the crushing phenomenon of the compressed concrete. Note that both kinds of behavior (tension-type failure, or compression-type failure) are governed by pivot B, meaning that the ultimate limit behavior is mostly controlled by the strain capacity of concrete in the presence of normal forces (this is, of course, obvious in case of steel reinforcements following an elastic and perfectly plastic behavior without any strain limitation in the steel reinforcements, but the same remark also holds in the presence of plasticity hardening). At the balanced failure point, as already mentioned in this chapter, the strain in the tensile steel reinforcement is equal to the maximum elastic strain in tension $\varepsilon_{s1} = \varepsilon_{su}$.

In Figure 1.24, the beam in pure bending ($N = 0$) fails due to the crushing of the concrete at the top of the beam, which means that pivot B controls the ULS of the bending behavior of this reinforced concrete section. However, pivot A failure can also be potentially observed with a strain of the tensile steel reinforcement that has reached the ultimate strain of steel in tension.

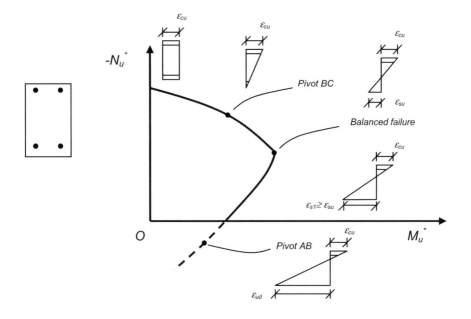

Figure 1.24. *Typical bending-compression interaction diagram – N ≤ 0 in compression; pivot B behaviour for N = 0 (bending case)*

As shown in Figure 1.24, the presence of a moderate compressive load in the "tension-type failure" branch increases the ultimate moment resistance of the section. The balanced failure point is the farthest right point on the interaction diagram and is also the point of the geometrical singularity of this interaction curve. Above the balanced failure point, in the "compression-type failure" branch, the presence of a significant compressive load decreases the ultimate moment resistance of the cross-section. This also implies less ductile behavior of the reinforced concrete member. When continuing to increase the amount of axial forces, the cross-section then behaves in pivot C (the transition between pivot B and pivot C is characterized by pivot BC) with the neutral axis position that is outside the cross-section. The ultimate point in Figure 1.24 is obtained in pure compression with uniform strain distribution.

The analytical and numerical interaction diagrams for symmetrical and unsymmetrical reinforced concrete sections have been studied by many authors. Some bending-curvature relationship and normal force-bending moment interaction diagrams were numerically obtained for symmetrically reinforced concrete beams by Pfrang *et al.* [PFR 64]. The sections have shown large amount of ductility at low axial load level. As the load level increased, ductility decreased markedly. Furthermore, the reinforced concrete beam became stiffer with increasing axial load up to a critical load called the balance load. To cite a few authors, we mention the books of Robinson [ROB 74], Park and Paulay [PAR 75], Lenschow [LEN 81], György [GYÖ 88], [MOS 07] and, more recently, Chandrasekaran *et al.* [CHA 09e] for the derivation of the analytical or numerical interaction diagrams (ultimate normal force–bending moment interaction diagram,).

For a rectangular cross-section, the dimensionless parameters can be defined from the ultimate internal forces calculated with respect to the center of gravity of the lower tensile steel reinforcements.

$$
\begin{cases}
\eta_u^* = \dfrac{N_u^*}{-bdf_{cu}} \\[4mm]
\mu_u^* = \dfrac{M_u^*}{-bd^2 f_{cu}}
\end{cases}
\text{and}
\begin{cases}
n = \dfrac{N_u^*}{bhf_{cu}} = -\dfrac{d}{h}\eta_u^* \\[4mm]
m = \dfrac{M_u^*}{-bh^2 f_{cu}} = \left(\dfrac{d}{h}\right)^2 \mu_u^*
\end{cases}
\qquad [1.284]
$$

In the next section, we will derive the closed-form expression of the interaction diagram characterized by the dimensionless normal force variable η_u^* and the dimensionless bending moment variable μ_u^*. We are essentially interested by the behavior in the compression regime for $N_u^* \le 0$, associated with the dimensionless normal force parameters η_u^* that is negative ($\eta_u^* \le 0$), or n that is positive ($n \ge 0$).

The dimensionless ultimate resistance variables can then be presented as:

$$
\begin{cases}
\eta_u^*(\alpha) = -\alpha\psi - \rho_{s1}' \dfrac{\sigma_{s1}}{f_{cu}} - \rho_{s2}' \dfrac{\sigma_{s2}}{f_{cu}} \\[4mm]
\mu_u^*(\alpha) = \mu(\alpha) - \dfrac{\alpha\psi}{2}(1-\delta') - \dfrac{\rho_{s1}'}{2}\dfrac{\sigma_{s1}}{f_{cu}}(1-\delta') + \dfrac{\rho_{s2}'}{2}\dfrac{\sigma_{s2}}{f_{cu}}(1-\delta')
\end{cases}
[1.285]
$$

where $\rho_{s1}' = A_{s1}/bd$ and $\rho_{s2}' = A_{s2}/bd$. This formula is general and valid for each concrete model (rectangular simplified law, parabola–rectangle law, bilinear law, etc.). In case of symmetrically reinforced concrete section, we have the steel area equality $\rho'_{s1} = \rho'_{s2} = \rho'_s/2$. It is possible to use also the dimensionless steel ratio ω expressed by $\omega = \rho'_s f_{su} /(-f_{cu})$.

1.3.1.2. *Analytical formulation of the interaction diagram*

1.3.1.2.1. Fundamental assumptions

We consider a symmetrically rectangular reinforced concrete section with both lower and upper steel reinforcements. The steel reinforcements are modeled with an elastic and perfectly plastic law (plasticity hardening is neglected). It clearly means that the ultimate cross-sectional behavior at the ULS is only controlled by pivot B, that is by the strain capacity of the compressed concrete. As a simplification, the concrete law is modeled by the simplified rectangular constitutive law for concrete in pivot B. However, we will also show that the parabola–rectangle constitutive law leads to very close results in pivot B, therefore legitimating the use of the simplified rectangular diagram. In pivot C, as also noticed by Robinson [ROB 74], Park and Paulay [PAR 75] or György [GYÖ 88], for instance, the simplified rectangular concrete law cannot be used anymore, and we will use the parabola–rectangle constitutive law.

The interaction diagram is known in each region of interest and the global interaction diagram is then obtained by piecing all solutions together. The boundary parameters α_i of each interval, where the interaction diagram is built, are given as:

$$\alpha'_{s2} = \frac{-\delta' \varepsilon_{cu3}}{-\varepsilon_{cu3} + \varepsilon_{su}}, \alpha_{s2} = \frac{-\delta' \varepsilon_{cu3}}{-\varepsilon_{cu3} - \varepsilon_{su}}, \alpha_{bal} = \frac{-\varepsilon_{cu3}}{-\varepsilon_{cu3} + \varepsilon_{su}},$$

$$\alpha_{BC} = \frac{h}{d} = 1 + \delta' \text{ and } \alpha_{s1} = \frac{-\varepsilon_{cu3}}{-\varepsilon_{cu3} - \varepsilon_{su}} \qquad [1.286]$$

When considering the parabola–rectangle constitutive law, subscript 3 has to be replaced by subscript 2 in the characteristic strains of concrete (for instance, ε_{cu3} by ε_{cu3}). The meaning of each characteristic parameter α_i is now analyzed while increasing the value of the dimensionless neutral axis position. $\alpha = \alpha'_{s2}$ characterizes the cross-sectional behavior where the upper steel reinforcement works in tension from the plastic to the elastic regime. $\alpha = \alpha_{s2}$ characterizes the cross-sectional behavior where the upper steel reinforcement works in compression from the elastic to the plastic regime.

$\alpha = \alpha_{bal}$ characterizes the cross-sectional behavior where the lower steel reinforcement works in tension from the plastic to the elastic regime. $\alpha = \alpha_{s1}$ is the position of the neutral axis from which both the lower and upper steel reinforcements work in compression in the plastic regime. When considering a B500B steel with $\varepsilon_{su} = 2.17 ‰$ coupled with the geometrical parameter $\delta' = 0.1$, the numerical values are obtained from equation [1.286]:

$$\alpha'_{s2} = 0.0617, \alpha_{s2} = 0.264, \alpha_{bal} = 0.617, \alpha_{BC} = 1.1 \text{ and } \alpha_{s1} = 2.64 [1.287]$$

When classifying the characteristic values α_i listed in equation [1.286], it appears, for instance, that it is mathematically quite obvious that $\alpha'_{s2} \leq \alpha_{s2}$, which physically expresses the fact that the upper steel reinforcements first work in tension and then in compression for an increasing value of α. This is, of course, confirmed by the numerical values presented in equation [1.287]. However, generally speaking, the comparison between $\alpha_{s2} = 0.264$ and $\alpha_{bal} = 0.617$ depends on the parameters of the cross-section as:

$$\alpha_{s2} \leq \alpha_{bal} \Rightarrow \delta' \leq \frac{-\varepsilon_{cu3} - \varepsilon_{su}}{-\varepsilon_{cu3} + \varepsilon_{su}} = 0.234 \text{ for } \varepsilon_{su} = 2.17‰ \qquad [1.288]$$

At the balanced failure point, the compression steel reinforcements are generally in the plasticity regime as outlined by Perchat [PER 09] for instance. Clearly for a B500B, we generally have $\delta' \leq 0.234$ (in the present case, we chose $\delta' = 0.1$). Each solution is now analytically detailed, and the global interaction diagram is finally obtained by piecing all these solutions together.

1.3.1.2.2. Pivot B – Steel reinforcements yielding in tension

This solution is valid for $\alpha \in [0; \alpha'_{s2}]$. For this range of parameters, the reinforced concrete section behaves in pivot B, and the lower and upper steel reinforcements yield in tension $(\sigma_{s1} = f_{su}, \sigma_{s2} = f_{su})$. Introducing these stress parameters in equation [1.285] leads to the ultimate capacity of the cross-section for this range of parameters:

$$\begin{cases} \eta_u^*(\alpha) = -\alpha \psi_B + \omega \\ \mu_u^*(\alpha) = \mu_B(\alpha) - \dfrac{\alpha \psi_B}{2}(1 - \delta') \end{cases} \qquad [1.289]$$

When considering, for instance, the parabola–rectangle constitutive law, this system of equation is specialized with:

$$\begin{cases} -\alpha\psi_B = -\dfrac{17}{21}\alpha \\ \mu_B(\alpha) - \dfrac{\alpha\psi_B}{2}(1-\delta') = \dfrac{17}{21}\alpha\left(\dfrac{1+\delta'}{2} - \dfrac{99}{238}\alpha\right) \end{cases} \qquad [1.290]$$

When introducing the simplified rectangular constitutive law for the concrete block, this equation is slightly changed in:

$$\begin{cases} -\alpha\psi_B = -0.8\alpha \\ \mu_B(\alpha) - \dfrac{\alpha\psi_B}{2}(1-\delta') = 0.8\alpha\left(\dfrac{1+\delta'}{2} - 0.4\alpha\right) \end{cases} \qquad [1.291]$$

In this last case, for the rectangular simplified concrete law, the parameterized system of equations is, in fact, equivalent to a parabolic relationship expressed in the plane $\left(\eta_u^*, \mu_u^*\right)$:

$$\mu_u^* = -\dfrac{\eta_u^{*2}}{2} + \eta_u^*\left(\omega - \dfrac{1+\delta'}{2}\right) + \dfrac{\omega}{2}(1+\delta' - \omega) \qquad [1.292]$$

1.3.1.2.3. Pivot B – upper steel reinforcement in elasticity

This solution is valid for $\alpha \in [\alpha_{s2}'; \alpha_{s2}]$. For this range of parameters, the reinforced concrete section behaves in pivot B. The lower steel reinforcements yield in tension and the upper steel reinforcement behaves in the elasticity range, in tension or in compression $(\sigma_{s1} = f_{su}, \sigma_{s2} = E_s\varepsilon_{s2} = E_s\varepsilon_{cu3}(\alpha-\delta')/\alpha)$. Introducing these stress parameters in equation [1.285] leads to the ultimate capacity of the cross-section for this range of parameters:

$$\begin{cases} \eta_u^*(\alpha) = -\alpha\psi_B + \dfrac{\omega}{2} + \dfrac{\omega}{2}\dfrac{\varepsilon_{cu3}}{\varepsilon_{su}}\dfrac{\alpha-\delta'}{\alpha} \\ \mu_u^*(\alpha) = \mu_B(\alpha) - \dfrac{\alpha\psi_B}{2}(1-\delta') + \dfrac{\omega}{4}(1-\delta') + \dfrac{\omega}{4}\dfrac{\varepsilon_{cu3}}{\varepsilon_{su}}\dfrac{\alpha-\delta'}{\alpha}(-1+\delta') \end{cases} \quad [1.293]$$

For the simplified rectangular constitutive law for the concrete block, the normal force equation allows us to identify the position of the neutral axis with respect to the normal force as:

$$\alpha^2 - \frac{5}{4}\alpha\left[\frac{\omega}{2}\left(1+\frac{\varepsilon_{cu3}}{\varepsilon_{su}}\right)-\eta_u^*\right]+\frac{5}{8}\delta'\omega\frac{\varepsilon_{cu3}}{\varepsilon_{su}}=0 \qquad [1.294]$$

whose solution of interest is:

$$\alpha\left(\eta_u^*\right)=\frac{5}{8}\left[\frac{\omega}{2}\left(1+\frac{\varepsilon_{cu3}}{\varepsilon_{su}}\right)-\eta_u^*\right]+\frac{5}{8}\sqrt{\left[\frac{\omega}{2}\left(1+\frac{\varepsilon_{cu3}}{\varepsilon_{su}}\right)-\eta_u^*\right]^2-\frac{8}{5}\delta'\omega\frac{\varepsilon_{cu3}}{\varepsilon_{su}}} \qquad [1.295]$$

For this range of parameters, the parameterized equation is a complex nonlinear function expressed in the plane $\left(\eta_u^*,\ \mu_u^*\right)$:

$$\mu_u^* = \frac{4}{5}\alpha\left(\frac{1+\delta'}{2}-\frac{2}{5}\alpha\right)+\frac{\omega}{4}(1-\delta')-\frac{\omega\,\varepsilon_{cu3}}{4\,\varepsilon_{su}}\frac{\alpha-\delta'}{\alpha}(1-\delta') \qquad [1.296]$$

where $\alpha\left(\eta_u^*\right)$ is extracted from equation [1.295].

1.3.1.2.4. Pivot B – steel reinforcements yielding in tension for the lower bars and in compression for the upper bars

This solution is valid for $\alpha\in[\alpha_{s2};\ \alpha_{bal}]$. For this range of parameters, the reinforced concrete section still behaves in pivot B. The lower steel reinforcements yield in tension and the upper steel reinforcements also yield but in the compression domain $(\sigma_{s1}=f_{su},\sigma_{s2}=-f_{su})$. Introducing these stress parameters in equation [1.285] leads to the ultimate capacity of the cross-section for this range of parameters:

$$\begin{cases} \eta_u^*(\alpha)=-\alpha\psi_B \\ \mu_u^*(\alpha)=\mu_B(\alpha)-\frac{\alpha\psi_B}{2}(1-\delta')+\frac{\omega}{2}(1-\delta') \end{cases} \qquad [1.297]$$

For this range of parameters, the parameterized equation is equivalent to a parabolic function expressed in the plane (η_u^*, μ_u^*):

$$\mu_u^* = -\frac{\eta_u^{*2}}{2} - \eta_u^* \frac{1+\delta'}{2} + \frac{\omega}{2}(1-\delta') \qquad [1.298]$$

At the balanced failure point, the dimensionless normal force is equal to $\eta_u^* = -0.8\alpha_{bal} = -0.494$ for a B500B steel. This leads to $n_{bal} = -\eta_u^*/(1+\delta') = 0.8\alpha_{bal}/1.1 = 0.449 \approx 0.45$ when δ' is chosen equal to 0.1. When using a parabola–rectangle concrete law, the dimensionless normal force at the balanced failure point is equal to $\eta_u^* = -0.81\alpha_{bal} = -0.5$, which is also equivalent to $n_{bal} = -\eta_u^*/(1+\delta') = 0.81\alpha_{bal}/1.1 = 0.4545 \approx 0.45$ when δ' is chosen equal to 0.1. The same value $n = 0.45$ has been obtained by Perchat, also using a parabola–rectangle constitutive law [PER 09]. Mari and Hellesland [MAR 05] obtained $n_{bal} = -\eta_u^*/(1+\delta') = 0.8\alpha_{bal}/1.1111 = 0.444$ when δ' is chosen equal to 0.111=1/9.

For the rectangular simplified concrete law, the parabolic function $m = f(n)$ for this range of parameters can also be expressed with the dimensionless parameters introduced in equation [1.284] as:

$$m = \frac{n}{2}(1-n) + \frac{\Omega}{2}\frac{1-\delta'}{1+\delta'}$$

$$\text{with } \Omega = \frac{A_s}{bh}\left(\frac{f_{su}}{-f_{cu}}\right) = \frac{d}{h} \quad \omega = \frac{\omega}{1+\delta'} \qquad [1.299]$$

For the parabola–rectangle concrete law, the parabolic function $m = f(n)$ for this range of parameters is slightly different:

$$m = \frac{n}{2}(1-1.028n) + \frac{\Omega}{2}\frac{1-\delta'}{1+\delta'}$$

$$\text{with } \Omega = \frac{A_s}{bh}\left(\frac{f_{su}}{-f_{cu}}\right) = \frac{d}{h} \quad \omega = \frac{\omega}{1+\delta'} \qquad [1.300]$$

The coefficient 1.028 in this parabolic function comes from $297/17^2 \approx 1.028$. When using the simplified rectangular concrete law, the exact nonlinear function for $\delta' = 0.1111 = 1/9$ is given by:

$$m = \frac{n}{2}(1-n) + \frac{2}{5}\Omega \tag{1.301}$$

which has been obtained by Mari and Hellesland, for instance, [MAR 05].

1.3.1.2.5. Pivot B – lower steel reinforcement in the elasticity range

This solution is valid for $\alpha \in [\alpha_{bal}; \alpha_{BC}]$. For this range of parameters, the reinforced concrete section still behaves in pivot B. The lower steel reinforcements now behave elastically and the upper steel reinforcement yields in the compression domain $(\sigma_{s1} = E_s \varepsilon_{s1} = E_s(\alpha-1)\varepsilon_{cu3}/\alpha, \sigma_{s2} = -f_{su})$. Introducing these stress parameters in equation [1.206] leads to the ultimate capacity of the cross-section for this range of parameters:

$$\begin{cases} \eta_u^*(\alpha) = -\alpha\psi_B + \frac{\omega}{2}\frac{\alpha-1}{\alpha}\frac{\varepsilon_{cu3}}{\varepsilon_{su}} - \frac{\omega}{2} \\[2mm] \mu_u^*(\alpha) = \mu_B(\alpha) - \frac{\alpha\psi_B}{2}(1-\delta') + \frac{\omega}{4}(1-\delta')\frac{\alpha-1}{\alpha}\frac{\varepsilon_{cu3}}{\varepsilon_{su}} + \frac{\omega}{4}(1-\delta') \end{cases} \tag{1.302}$$

For the simplified rectangular constitutive law for the concrete block, the normal force equation allows us to identify the position of the neutral axis with respect to the normal force as:

$$\alpha^2 - \frac{5}{4}\alpha\left[\frac{\omega}{2}\left(-1 + \frac{\varepsilon_{cu3}}{\varepsilon_{su}}\right) - \eta_u^*\right] + \frac{5}{8}\omega\frac{\varepsilon_{cu3}}{\varepsilon_{su}} = 0 \tag{1.303}$$

whose solution of interest is:

$$\alpha(\eta_u^*) = \frac{5}{8}\left[\frac{\omega}{2}\left(-1 + \frac{\varepsilon_{cu3}}{\varepsilon_{su}}\right) - \eta_u^*\right]$$
$$+ \frac{5}{8}\sqrt{\left[\frac{\omega}{2}\left(-1 + \frac{\varepsilon_{cu3}}{\varepsilon_{su}}\right) - \eta_u^*\right]^2 - \frac{8}{5}\omega\frac{\varepsilon_{cu3}}{\varepsilon_{su}}} \tag{1.304}$$

For this range of parameters, the parameterized equation is a complex nonlinear function expressed in the plane (η_u^*, μ_u^*):

$$\mu_u^* = \frac{4}{5}\alpha\left(\frac{1+\delta'}{2} - \frac{2}{5}\alpha\right) + \frac{\omega}{4}(1-\delta')$$

$$+ \frac{\omega\,\varepsilon_{cu3}}{4\,\varepsilon_{su}}\frac{\alpha-1}{\alpha}(1-\delta') \qquad\qquad [1.305]$$

where $\alpha(\eta_u^*)$ is extracted from equation [1.294].

1.3.1.2.6. Pivot C – lower steel reinforcement in the elasticity range

In pivot C, as also noted by Robinson [ROB 74], Park and Paulay [PAR 75] or György [GYÖ 88], for instance, the simplified rectangular concrete law cannot be used anymore, and we will use the parabola–rectangle constitutive law.

This solution is valid for $\alpha \in [\alpha_{BC}; \alpha_{s1}]$. For this range of parameters, the reinforced concrete section behaves in pivot C. The lower steel reinforcements still behave elastically and the upper steel reinforcement yields in the compression domain $(\sigma_{s1} = E_s\varepsilon_{s1} = E_s(\alpha-1)\varepsilon_{cu3}/\alpha, \sigma_{s2} = -f_{su})$. Introducing these stress parameters in equation [1.285] leads to the ultimate capacity of the cross-section for this range of parameters:

$$\begin{cases} \eta_u^*(\alpha) = -\alpha\psi_C + \dfrac{\omega}{2}\dfrac{\alpha-1}{\alpha}\dfrac{\varepsilon_{cu3}}{\varepsilon_{su}} - \dfrac{\omega}{2} \\[2ex] \mu_u^*(\alpha) = \mu_C(\alpha) - \dfrac{\alpha\psi_C}{2}(1-\delta') + \dfrac{\omega}{4}(1-\delta')\dfrac{\alpha-1}{\alpha}\dfrac{\varepsilon_{cu3}}{\varepsilon_{su}} + \dfrac{\omega}{4}(1-\delta') \end{cases} \qquad [1.306]$$

where $\alpha\psi_C$ and μ_C are given by equations [3.156] and [3.157] of [CAS 12]. The parameterized equations of equation [1.296] describe a complex nonlinear function.

1.3.1.2.7. Pivot C – both steel reinforcements yield in compression

The last range of parameters is concerned by $\alpha \geq \alpha_{s1}$, where both the lower and upper steel reinforcements are in compression and behave in plasticity $(\sigma_{s1} = -f_{su}, \sigma_{s2} = -f_{su})$. Introducing these stress parameters in equation [1.285] leads to the ultimate capacity of the cross-section for this range of parameters:

$$
\begin{cases}
\eta_u^*(\alpha) = -\alpha\psi_C - \omega \\
\mu_u^*(\alpha) = \mu_C(\alpha) - \dfrac{\alpha\psi_C}{2}(1-\delta')
\end{cases}
\qquad [1.307]
$$

where $\alpha\psi_C$ and μ_C are given by equations [3.156] and [3.157] of [CAS 12]. For this range of parameters, the ultimate bending moment–normal force curve is linear (straight line in the interaction diagram) and is given by:

$$
\mu_u^* = \frac{1+\delta'}{2}(1+\delta'+\omega+\eta_u^*)\left(1-\frac{\varepsilon_{c2}}{2\varepsilon_{cu2}}\right)
\qquad [1.308]
$$

The case of pure compression is simply obtained by putting $\mu_u^* = 0$ in equation [1.308] giving $\eta_u^* = -1-\delta'-\omega$ or equivalently $n = 1+\Omega$. Equation [1.308] can also be written as:

$$
m = \frac{1}{2}(1+\Omega-n)\left(1-\frac{\varepsilon_{c2}}{2\varepsilon_{cu2}}\right)
\qquad [1.309]
$$

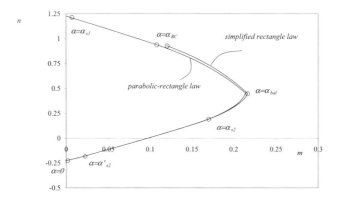

Figure 1.25. *Comparison of the interaction diagram obtained for a symmetrically reinforced concrete section for two concrete laws: simplified rectangular law and parabolic-rectangle law;*
$\omega = 0.25; \ \delta' = 0.1 \ and \ f_{su} = 500 \ MPa$

Figure 1.25 shows a typical interaction diagram for a symmetrically reinforced concrete rectangular section. The simplified rectangular law gives a good approximation of the exact interaction diagram derived with the parabola–rectangle law for concrete, at least up to pivot BC. The sharp angle at the balanced failure point is highlighted for both interaction diagrams at almost the same location.

1.3.2. *Approximation formulations*

From an engineering point of view, the exact interaction diagram is not easy to manipulate for fast and robust designs (even in the case of the simplified rectangular diagram), and approximation formula expressed in a simpler way could be required for engineering purposes. One of the simplest approximations is based on a bilinear approximation of the complex nonlinear boundary of the interaction diagram, where the balanced failure point serves as one of the reference points for building both straight lines (see, for instance, [PER 09]). The straight line compression failure above the balance failure point is a known approximation as reported, for instance, by Park and Paulay [PAR 75]. This bilinear interaction curve can be decomposed into two branches, around the balanced failure point characterized by:

$$n_{bal} = \frac{\alpha_{s2}\psi_B}{1+\delta'} \; ; m_{bal} = \frac{n_{bal}(1-n_{bal})}{2} + \frac{\Omega}{2}\frac{1-\delta'}{1+\delta'} \qquad [1.310]$$

Using B500B steel and the simplified rectangular diagram ($\psi_B = 0.8$), we obtain $n_{bal} = 0.449$ for $\delta' = 0.1$, and $n_{bal} = 0.444$ for $\delta' = 0.111$. A typical value for n_{bal} is between 0.40 and 0.45 (see [PER 09] for instance). The equations of the bilinear approximation are then expressed by:

$$\begin{cases} n \leq n_{bal} \Rightarrow m = \left(1 - \dfrac{m_0}{m_{bal}}\right)n + m_{bal} \\[3mm] n \geq n_{bal} \Rightarrow m = \dfrac{m_{bal}}{n_{bal} - (1+\Omega)}[n - (1+\Omega)] \end{cases} \qquad [1.311]$$

where m_0 is the ultimate limit moment in simple bending (without normal force coupling $n=0$). A comparison between the bilinear approximation and the exact interaction diagram based on a parabola–rectangle law for concrete is shown in Figure 1.26.

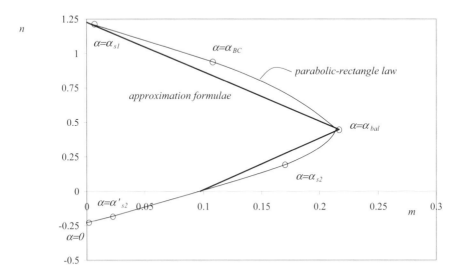

Figure 1.26. *Interaction diagram for a symmetrically reinforced concrete section; comparison of the bilinear approximation with the exact boundary;*
$\omega = 0.25; \quad \delta' = 0.1 \quad and \quad f_{su} = 500 \quad MPa$

The bilinear approximation is simple but is not safe, as the exact interaction diagram is strongly nonlinear (see Figure 1.26). Mari and Hellesland improved the bilinear approximation by adding a parabolic nonlinear law below the balanced failure point that has been analytically derived [MAR 05]. A second characteristic point is used for the construction of the approximation diagram now based on three branches:

$$n_{s2} = \frac{\alpha_{s2}\psi_B}{1+\delta'}; m_{s2} = \frac{n_{s2}(1-n_{s2})}{2} + \frac{\Omega}{2}\frac{1-\delta'}{1+\delta'} \qquad [1.312]$$

Using B500B steel and the simplified rectangular diagram ($\psi_B = 0.8$), we obtain $n_{s2} = 0.192$ for $\delta' = 0.1$, and $n_{s2} = 0.211$ for $\delta' = 0.111$. A typical value reported by Mari and Hellesland is $n_{s2} = 0.2$ [MAR 05]. The equations of the nonlinear approximation proposed by Mari and Hellesland are then expressed by [MAR 05]:

$$\begin{cases} n \leq n_{s2} \Rightarrow m = \left(\dfrac{1}{2} - n_{s2}\right)(n - n_{s2}) + m_{s2} \\[2mm] n \in [n_{s2}; n_{bal}] \Rightarrow m = \dfrac{n}{2}(1-n) + \dfrac{\Omega}{2}\dfrac{1-\delta'}{1+\delta'} \\[2mm] n \geq n_{bal} \Rightarrow m = \dfrac{m_{bal}}{n_{bal} - (1+\Omega)}[n - (1+\Omega)] \end{cases} \qquad [1.313]$$

In this method, the first branch is computed from the continuity of the tangent at the point parameterized by $\alpha = \alpha_{s2}$. This method gives better results especially below the balanced failure point (see Figure 1.27).

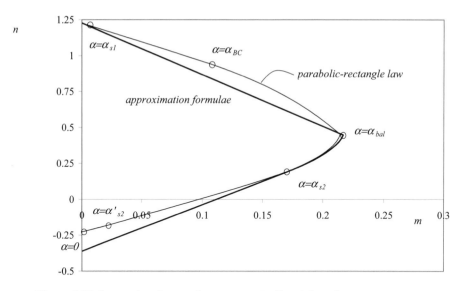

Figure 1.27. *Interaction diagram for a symmetrically reinforced concrete section; comparison of the approximation formula of Mari and Hellesland [MAR 03, MAR 05] with the exact boundary; $\omega = 0.25$; $\delta' = 0.1$ and $f_{su} = 500$ MPa*

A better (but more complex) approximation would be based on a modified first branch for $n \leq n_{s2}$, by connecting straight line to the point in pure tension $m=0$ for which $n=-\Omega$. Furthermore, a third characteristic point is used for the construction of the approximation diagram now based on four branches:

$$n_{BC} = \psi_B - \frac{\Omega}{2}\left(\frac{\delta'}{1+\delta'}\frac{\varepsilon_{cu3}}{\varepsilon_{su}} - 1\right); m_{BC} = \frac{5}{14}(1+\Omega - n_{BC}) \qquad [1.314]$$

It is assumed that the tangent is continuous at the connection of pivot BC. It would be possible to use an approximation formula based on four branches as:

$$\begin{cases} n \le n_{s2} \Rightarrow m = m_{s2}\dfrac{n+\Omega}{n_{s2}+\Omega} \\[2mm] n \in [n_{s2}; n_{bal}] \Rightarrow m = \dfrac{n}{2}(1-n) + \dfrac{\Omega}{2}\dfrac{1-\delta'}{1+\delta'} \quad \text{with} \\[2mm] n \in [n_{bal}; n_{BC}] \Rightarrow m = a_1 n^2 + a_2 n + a_3 \\[2mm] n \ge n_{BC} \Rightarrow m = \dfrac{5}{14}(1+\Omega - n) \end{cases}$$

$$\begin{cases} a_1 = \dfrac{m_{bal} - m_{BC} + \dfrac{5}{14}(n_{bal} - n_{BC})}{(n_{bal} - n_{BC})^2} \\[3mm] a_2 = -\dfrac{5}{14} - 2a_1 n_{BC} \\[2mm] a_3 = m_{BC} - a_1 n_{BC}^2 - a_2 n_{BC} \end{cases} \qquad [1.315]$$

Figure 1.28 shows the efficiency of this four-branch interaction diagram that is composed of two straight lines and two parabolic branches connected at the balanced failure point.

The simplified interaction diagrams of symmetrically reinforced concrete sections in two opposite faces normal to the plane of bending (as treated in this chapter), equally reinforced in the four faces, or symmetrically reinforced in two opposite faces parallel to the plane of bending, are analytically given in Mari and Hellesland [MAR 05], based on a three-branch diagram with two straight lines and one parabolic branch (see Figure 1.27 for the analysis of this method).

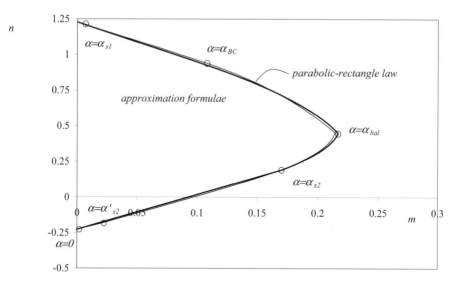

Figure 1.28. *Interaction diagram for a symmetrically reinforced concrete section; comparison of a four branches-based approximation formula with the exact boundary; $\omega = 0.25;\ \delta' = 0.1\ and\ f_{su} = 500\ MPa$*

It is quite interesting to note that interaction diagrams are not explicitly given in Eurocode 2 (reinforced concrete structures) but are used in Eurocode 4 (steel-concrete composite structures for instance). Eurocode 4 suggests, for instance, a tri-linear normalized interaction diagram, whereas Perchat [PER 09] introduced a bilinear interaction diagram. As already detailed, parabolic branches can be added for more accurate and safer design.

1.3.3. *Graphical results for general cross-sections*

The interaction diagrams are sensitive to the reinforcement density, but also to the shape of the cross-section. In Figures 1.29–1.32, the ultimate normal force–bending moment interaction diagram of a reinforced cross-section (with bilinear law for concrete in compression and for steel) are presented for different amounts of steel ratio Ω. Hardening effect is neglected for the steel behavior (perfect plasticity for steel without limitation of the steel strain, i.e. no pivot *A* at ULS). These diagrams have been computed in the book *Design Aids for EC2* [DES 05].

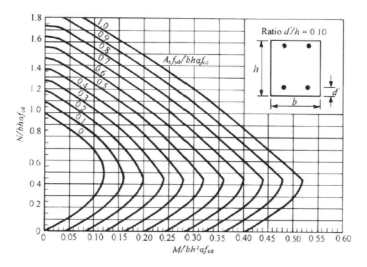

Figure 1.29. *Bending-compression Interaction diagram for a rectangular symmetrically reinforced concrete section (after [DES 05]); d'/h = 0.1; α = 0.85*

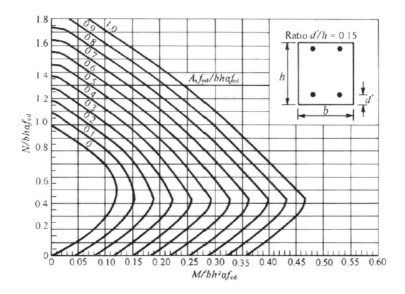

Figure 1.30. *Bending-compression interaction diagram for a rectangular symmetrically reinforced concrete section (after [DES 05]); d'/h = 0.15; α = 0.85*

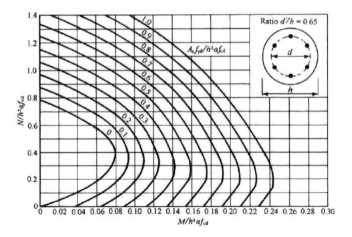

Figure 1.31. *Bending-compression interaction diagram for a circular symmetrically reinforced concrete section (after [DES 05]); d/h = 0.65; α = 0.85*

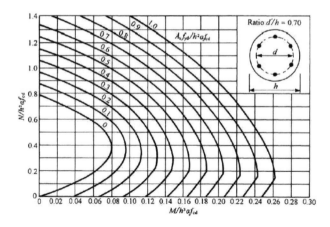

Figure 1.32. *Bending-compression interaction diagram for a circular symmetrically reinforced concrete section (after [DES 05]); d/h = 0.70; α = 0.85*

The interaction diagrams are shown below for different kind of reinforced concrete sections. In Figures 1.29–1.32, the normal force is counted positive in compression. The interaction diagrams can be directly read with the dimensionless variables (n,m) introduced above. The strength characteristic is decreased by the dimensionless coefficient $α = 0.85$ for these calculations. It should be noted that the product "$αf_{cd}$", in the dimensionless moment and axial force coordinates of the diagrams, is identical to the design compressive strength "f_{cd}" in EC2. The normal force and bending moment are expressed with respect to the center of gravity of the concrete section. It is shown that the geometrical singularity of the interaction diagram of a rectangular reinforced cross-section is less pronounced for a circular cross-section.

An example of a interaction diagram valid for an unsymmetrical reinforced concrete section is given in Figure 1.33. This rectangular cross-section is reinforced by some lower and upper steel reinforcements ($6\phi20$ for the lower part and $3\phi20$ for the upper part). The reinforcements are made of B400B steel with f_{yk} = 400 MPa (f_{su} = 347.8 MPa). The hardening effect is included for the steel modeling and the hardening parameter k is equal to k = 1.08 (steel with high ductility). The concrete used in this reinforced concrete section is a C40/45 type with f_{ck} = 40 MPa (f_{cu} = −26.67 MPa). The concrete is modeled by the parabolic rectangular law at ULS. The normal force and bending moments at ULS are also calculated with respect to the center of gravity of the rectangular concrete section. The balanced failure points appear at both sides of the interaction diagram for positive and negative bending moments. However, these balanced failure points are not symmetrical with respect to the vertical axis, that is they are not associated with the same ultimate normal force level (due to the unsymmetrical reinforcements of this reinforced concrete section).

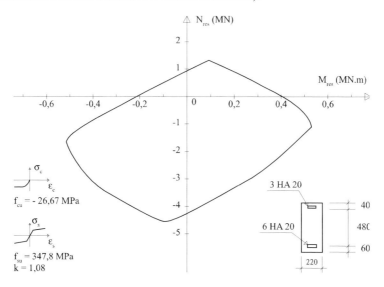

Figure 1.33. *Interaction diagram for an unsymmetrical cross-section; N ≤0 in compression*

It is numerically observed for this reinforced concrete section that the section behaves in pivot B for $N = 0$. Pivot A controls the failure of the reinforced concrete section, especially in the tension regime of course. It is worth mentioning that the interaction diagram is not symmetric in this space due to the loss of symmetry induced by the unsymmetrical reinforcements of this rectangular cross-section.

Chapter 2

Slender Compression Members – Mechanics and Design

2.1. Introduction

In structures that include slender compression members, it is necessary to consider second-order load effects in the analysis and design. In this chapter, general principles of analysis and design of such structures, and individual elements, are discussed to provide a general understanding of the problem area. An effort is made to provide a reasonable complete basis for relevant analysis and design requirements and recommendations in Eurocode 2:2004 code (for convenience, labeled EC2 in the following). Also, some additional materials that may be found useful for students, practicing engineers and researchers are presented.

2.2. Analysis methods

2.2.1. *General*

A fully nonlinear analysis of a reinforced concrete structure should include effects of both nonlinear material properties of the constituent materials (concrete and reinforcing steel) and geometrically nonlinear effects (i.e. the effect of axial forces on deformations). Such analyses are today generally based on finite element method (FEM) formulations. For building structures, it is normally acceptable to base calculations on small deflection (rotation) theory. In that case, geometrically nonlinear effects can be accounted for through second-order theory.

Geometrically nonlinear effects may be global or local in nature. In the first case, they may, for instance, be due to increased lateral deflections of the structure as a whole, caused by the gravity loading on the sidesway of the overall structure. In the second case, they will be due to axial forces acting on the deflection between ends of individual members, thereby causing increased member deflections and moments. Clearly, there is an interaction between the two "types", but, nevertheless, this classification is often found convenient.

In elastic second-order theory, nonlinear effects will be solely due to the action of axial forces on the deflected structure and individual members. Material nonlinearity will, in such theories, have to be taken into account in simplified ways through reduced, secant stiffness assumptions. In simplified theories, first-order and second-order effects are calculated separately.

The basic requirements in structural mechanics analysis are satisfied to different extents in different theories:

1) Full nonlinear analysis will satisfy all three basic requirements: i) equilibrium (at a significant number of selected sections); ii) kinematic compatibility and iii) constituent material laws (of concrete, reinforcing steel). Such theories are not considered here.

2) Second-order elastic theory, with reduced stiffness to reflect cracking, nonlinear material laws, time dependent effects (creep, shrinkage), can, in principle, be made to satisfy equilibrium and kinematic compatibility (between external and internal forces and curvatures) at any number of sections. However, the section check will often be limited to the critical section of a member. Although second-order elastic theory is dealt with superficially in EC2, it will be presented and discussed in more detail in this chapter.

3) Approximate methods, based on load effects from a combination of first-order elastic analysis and subsequent second-order corrections, consider equilibrium between external forces and internal resistances at the critical section of a member only. Compatibility is tacitly considered satisfied in such methods. EC2 gives two methods of this type (the nominal stiffness and the nominal curvature method), and they are between the most commonly used methods in practical design work. Much of this chapter aims at providing a reasonable good background for such methods.

2.2.2. *Requirements to second-order analysis*

General principles (P) for second-order analysis are given in EC2 (5.8.2). They are for the convenience of referencing quoted verbatim below:

1) P This clause deals with members and structures in which the structural behavior is significantly influenced by second-order effects (e.g. columns, walls, piles, arches and shells). Global second-order effects are likely to occur in structures with a flexible bracing system.

2) P Where second-order effects are taken into account (see point 6), equilibrium and resistance shall be verified in the deformed state. Deformations shall be calculated by taking into account the relevant effects of cracking, nonlinear material properties and creep.

Note. In an analysis assuming linear material properties, this can be taken into account by means of reduced stiffness values, see 5.8.7.

3) P Wherever relevant, analysis shall include the effect of flexibility of adjacent members and foundations (soil-structure interaction).

4) P The structural behavior shall be considered in the direction in which deformations can occur, and biaxial bending shall be taken into account when necessary.

5) P Uncertainties in geometry and position of axial loads shall be taken into account as additional first-order effects based on geometric imperfections, see 5.2.

6) P Second-order effects may be ignored if they are less than 10% of the corresponding first-order effects. Simplified criteria are given for isolated members in 5.8.3.1 and for structures in 5.8.3.3.

2.3. Member and system instability

2.3.1. *Elastic critical load and effective (buckling) length*

Before discussing specific methods, it will be useful for the understanding of these methods to review the concepts of critical load and effective length, or buckling length, for some simple linear elastic compressions members, and place these in relation to a reinforced concrete member.

The load–deflection relationship of an initially straight compression member is shown by the bilinear curve with branches (a) and (b) in Figure 2.1. The member remains straight (a) till the critical load N_{cr} (labeled N_B in EC2, and used with that notation later) is reached. A little disturbance of the member as the top of branch (a) is reached, will cause it to bifurcate (deflect sideways) toward the right or left. In second-order theory, this deflection will take place without any change in load, as reflected by the horizontal deflection line (b) in the figure. In the accurate, large deflection theory, a slight increase in load is required to increase the sideways deflection as indicated by the broken line (c) in the figure.

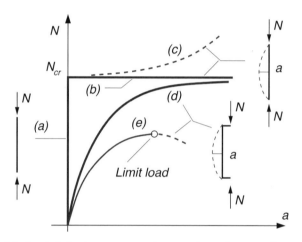

Figure 2.1. *Load–deflection response of: (a) perfect column for $N < N_{cr}$, (b) perfect column for $N = N_{cr}$ according to small deflection theory, (c) perfect column for $N > N_{cr}$ according to large deflection theory, (d) imperfect elastic column and (e) imperfect reinforced concrete column*

If the member was not initially straight, but had, for instance, an imperfection, or was subjected to some transverse loading or end moments, the member will gradually deflect as indicated by the curve labeled (d) in the figure, and approach asymptotically the horizontal line as the axial load approaches the critical load. Case (e) is for a column with nonlinear material properties, and will be discussed in conjunction with Figure 2.3.

The analysis of an imperfect column represents a second-order bending (flexural) problem. The bifurcation problem of a perfect column, however, is a classical buckling or eigenvalue problem. The deflection shape (buckling mode) at the critical load can be determined, but the deflection magnitude remains unknown.

The theoretical buckling or bifurcation load of the perfect elastic column is a useful parameter, also in the formulation of simplified second-order methods for columns with nonlinear material properties.

Buckling modes for some columns with constant sectional bending stiffness EI and axial force are shown in Figure 2.2. The distance between inflection points (points of contraflexure at which moments are zero) is normally called the elastic buckling or effective length, and in EC2 it is denoted by L_0. If the inflection points are located at ends (Case a) or between column ends (Case b), L_0 will be equal to or smaller than the column length L. If one or more inflection points are located outside the column length, on the mathematical continuation of the buckled curve, as in Cases (c) and (d), the effective length will be greater than the member length.

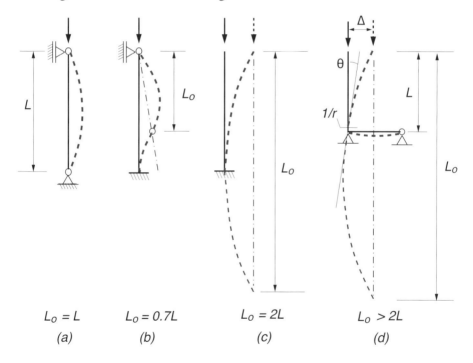

$L_0 = L$ $L_0 = 0.7L$ $L_0 = 2L$ $L_0 > 2L$

(a)   (b)   (c)   (d)

Figure 2.2. *Effective length examples: (a) pinned-end column, (b) pinned-fixed column, (c) cantilever column fixed at the base, (d) flexibly restrained cantilever column*

The buckling curves have a sinusoidal variation along the member relative to the axial thrust line (though the inflection points). The effective length, L_0, is thus equal to half the wave length of this sine curve. Then, at a distance x from an inflection point, the load eccentricity and the moment are given by:

$$w = w_o \sin \frac{\pi x}{L_0} \qquad \text{and} \qquad M = -EIw'' = EIw_o \left(\frac{\pi}{L_0}\right)^2 \sin \frac{\pi x}{L_0}$$

where w_0 is the deflection at the mid–length of L_0. Further, L is the member length and EI the sectional bending stiffness. The critical (buckling or bifurcation) load, N_{cr}, can then be obtained from moment equilibrium at any section ($M = N_{cr}w$) as:

$$N_{cr} = \frac{M}{w} = \frac{\pi^2 EI}{(L_0)^2} = \frac{N_E}{\beta_0^2} \qquad\qquad\qquad [2.1]$$

where

$$L_0 = \beta_0 L \qquad \text{and} \qquad N_E = \frac{\pi^2 EI}{L^2} \qquad\qquad [2.2]$$

Here, β_0 is the effective length factor and N_E is the Euler load (i.e. the buckling load of a pinned-end member).

This is the well-known, classical critical (buckling or bifurcation) load expression. It is normally derived directly from the differential equation for the eigenvalue problem, and neglects effects of shear deformations. For members with varying stiffness and axial force along the length, EI and N are the values at a chosen reference section.

The above relationships can also be used to establish an expression for the deflection at an arbitrary distance in terms of the curvature at that distance. Consider Figure 2.2, Case (d). At the column base, a distance $x = L$ from the top end (taken as the origin), the top deflection Δ and curvature $1/r$ become:

$$\Delta = w(L) = w_o \sin \frac{\pi L}{L_0} \qquad \text{and} \qquad \frac{1}{r} = -w''(L) = \left(\frac{\pi}{L_0}\right)^2 w_o \sin \frac{\pi L}{L_0}$$

From these, a very useful relationship between the top deflection expressed by the curvature at $x = L$ can be derived as:

$$\Delta = \frac{L_0^2}{\pi^2} \cdot \frac{1}{r} \qquad\qquad [2.3]$$

This will be used later in the simplified "curvature method". The ratio L_0/h, where h is the cross–section depth (or height), is an often used measure of geometric slenderness of a compression member. But more often, λ is used. It can be defined by rewriting the classical, linear elastic buckling load, equation [2.1]:

$$N_{cr} = \frac{\pi^2 EI}{L_0^2} = \frac{\pi^2 EA}{\lambda^2} \qquad\qquad [2.4]$$

where

$$\lambda = L_0/i \qquad \text{and} \qquad i = \sqrt{EI/EA} = \sqrt{I/A} \qquad\qquad [2.5]$$

are the *mechanical slenderness* and radius of gyration, respectively, and EA is the axial stiffness of the section. For a homogeneous section, the elastic modulus E cancels out, leaving i to be defined by the second moment of area (about the center of area), I, and the area A. The slenderness defined by L_0/h, where h is the section height in the buckling direction, is generally termed *geometric slenderness* because it is a function of geometric properties only. However, because the mechanical property I also is a function of geometric properties, λ (equation [2.5]) is also sometimes referred to as geometric slenderness in the literature.

In conjunction with simplified methods for reinforced concrete members, to be considered later, the slenderness definition terms above are used with $A = A_c$ and $I = I_c$ for the area and second moment of area of the gross, uncracked concrete section, respectively. Typical values are:

$$\lambda = L_0/0.29h \qquad \text{and} \qquad \lambda = L_0/0.25h$$

for rectangular and circular cross-sections.

2.3.2. System instability principles

The critical loading at which a structure is on the verge of instability (buckling), can be obtained from system instability (buckling) analysis as given by the zero determinant condition of the system stiffness matrix.

In the common case of *proportional loading*, when all loads are increased by the same load factor λ_f, the critical loading for a given initial axial force (N) distribution in the members, can be defined through a single (critical) load factor λ_{fcr} Inversely, it may be defined through a single *system stability index*, $\alpha_{system} = 1/\lambda_{fcr}$, common to all members.

Alternatively, the critical system loading may be reflected in terms of the critical load and effective (buckling) length of any individual compression member. Then, when $N_{cr} = N/\alpha_{system}$ is known, the effective length factor becomes:

$$\beta_0 = \sqrt{\frac{N_E}{N_{cr}}} = \sqrt{\frac{\alpha_{system}}{\alpha_E}} \qquad \text{where} \qquad \alpha_E = \frac{N}{N_E} \qquad [2.6]$$

is a nominal member flexibility parameter, or "load index", for the given reference load. The relationship between the effective length factors of any two members is interrelated through the common system stability index α_{system}. A consequence of this is that the individual *member stability indices* become identical in all members:

$$\alpha_i = \frac{N_i}{N_{cr,i}} = \alpha_{Ei} \beta_{0i}^2 \qquad i = 1, 2, 3... \qquad [2.7]$$

2.3.3. Concrete column instability – limit load

For a reinforced concrete member, with nonlinear material properties, the load–deflection response might be something like the curve labeled (e) in Figure 2.1. It will be discussed more with reference to Figure 2.3.

Upon increasing the axial force, the column reaches a limit load (at the peak) as the member instability condition is reached. If a load increase had been attempted at this point, it would not be possible to establish equilibrium between external and internal resisting moments, and uncontrolled deflection would have resulted. However, if the column deflection had been imposed (controlled), and the corresponding axial load lowered to maintain equilibrium, an unstable softening, or descending, load–deflection branch might result, as indicated in the figure. For real framed columns, where the column might shed its load to neighboring elements, such response can be encountered.

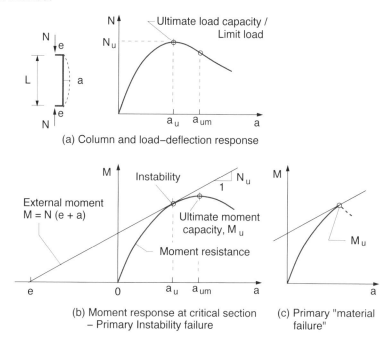

Figure 2.3. *RC column and section response and instability*

The limit load is obtained at a finite deflection, and is often also referred to as the ultimate load capacity, instability load or stability failure load. This load, denoted N_u in the figure, may be significantly smaller than the elastic buckling (bifurcation) load.

The load capacity of reinforced concrete members may be limited by primary stability failure or by primary "material failure". The first case is that illustrated in Figures 2.3(a) and (b). It is seen that such a failure takes place before the ultimate moment capacity (maximum moment resistance), M_u, of the critical (maximum moment) section is exhausted. The state represented by M_u is often referred to, somewhat imprecisely, as a material failure state.

M_u may, in some cases, be located on the ascending branch of the moment resisting curve (rather than at a peak with a horizontal tangent), at or below the instability point in Figure 2.3(b). Such a case is illustrated in Figure 2.3(c), in which the moment capacity M_u will typically result in the attainment of predefined failure strains in the constituent materials. As seen, M_u limits the load for which it will be possible to establish equilibrium in such cases. Loss of stability can, therefore, be said to be due to primary material failure.

2.4. First- and second-order load effects

2.4.1. *Global and local second-order effects*

Second-order effects due to axial loads affect interconnected members of a frame structure in three ways:

1) in an overall, or *global* frame sense, due to sidesway of the frame system as such;

2) in an individual, or *local* member sense, due to axial loads acting on the deflections away from the chord between member ends, and thereby giving rise to nonlinear (curved) moment distributions;

3) in a *local* sense, due to the effect of axial loads on rotational end restraint stiffnesses at member ends. In most cases, such effects are normally small, thereby allowing approximate analyses to be based on first-order end restraints stiffnesses.

However, the effect on rotational end restraint stiffnesses may be significant in some cases. For instance, in regions of multilevel columns (multistory frames) with strong interaction between columns on adjacent levels, which is typical for regions with columns in single curvature bending [HEL 09b]. And also due to strong horizontal interaction, through connecting beams, of columns with significant stiffness differences (due to different sections and axial force levels) in neighboring axes.

In approximate analyses, a conventional first-order analysis is carried out first. This analysis implicitly implies first-order end rotational restraints stiffnesses. Subsequently, second-order effects are considered. This subdivision into global and local effects is very common, and sometimes referred to as $N\Delta$ and $N\delta$ (or $P\Delta$ and $P\delta$) effects, respectively.

There is an interaction between global and local effects. Sway due to global effects affects the moment gradients along individual columns, and therefore the development of curved moment distributions and maximum moments between ends. Local effects (curved moment distributions) in individual, axially loaded columns affect the lateral displacement Δ and, consequently, the sidesway moments.

2.4.2. Single members

The difference between load effects computed by second-order and first-order theory is most often referred to as second-order or geometric nonlinear effects (or secondary effects).

The simple cantilever example in Figure 2.4(a) illustrates the differences for a statically determinate unbraced element. The second-order moment (M_2) is directly given by the axial load times the deflection. For the indeterminate unbraced example in Figure 2.4(b), the second-order moment (M_2) is directly given by the axial load times the deflection measured from the vertical chord through the inflection point. The first-order moment (M_0) diagrams are linear when no distributed, transverse loading is applied to the members. The second-order moment diagrams for the columns are curved due the curved deflection shapes, and the maximum first-order moment and the maximum second-order moment are located at the same section.

In the braced, rotationally restrained column case in Figure 2.5(a), the situation is a bit more complicated. Maximum first-order moment and maximum second-order moment are not located at the same section anymore. Also shown in the figure is the buckled shape and the corresponding moment diagram, Figure 2.5(c). It can be seen to be similar to the second-order moment diagram in Figure 2.5(b), obtained as the difference between the total moment and the first-order moment.

An estimate of the total moment distribution along a statically indeterminate column can be obtained by adding the buckling moment (with an appropriate maximum value) to the first-order diagram. EC2, and other codes, suggests this as an approach for determining the maximum moment along the column length. The implication of this is that the inflection points in the two cases coincide approximately so that the effective length is approximately the same as the distance between the intersection points between the total and the first-order moment diagrams in Figure 2.5(b). In a study of this and other aspects, it has been pointed out [LAI 83b] that there may be significant differences between these lengths. Nevertheless, the summation approach does provide an approximate total moment distribution. Practical approaches for determining maximum total moment will be covered later.

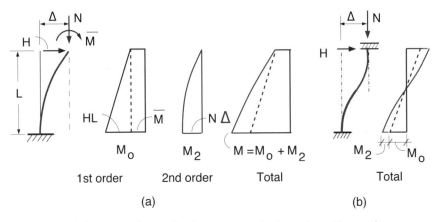

Figure 2.4. *First- and second-order moments of unbraced, cantilever columns*

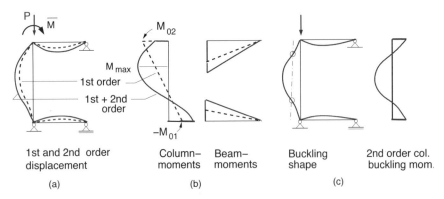

Figure 2.5. *First- and second-order column and beam moments of a simple, braced frame*

2.4.3. *Frame mechanics – braced and bracing columns*

2.4.3.1. *General*

Structures and compression members are defined as *braced* or *unbraced* depending upon whether the overall lateral stability is provided by some kind of external bracing system (lateral supporting device, shear walls, truss, etc.), or not.

The term "braced" is normally used also to denote a frame or column of which the lateral displacement is completely prevented. However, this is a rather academic case. Indeed, for practical bracing stiffnesses, there will normally be some lateral displacement. The term "braced" does not *per se* exclude some limited lateral displacement. Therefore, when there is a need for more precision, the term *partially braced* may be used. Similarly, the term *fully braced* can be used to denote a frame or column of which the lateral displacement is completely prevented.

In an unbraced frame with stiff and flexible columns interconnected by beams, floors, etc., we may have both "braced" and "bracing" (unbraced) columns. For lateral stability of such frames, the stiffer columns will interact with, and provide lateral bracing to, the more flexible columns, that without this bracing might become laterally unstable. Such columns can be considered braced at an imposed lateral sway displacement, restricted to that of the frame.

These cases are illustrated in Figure 2.6 for an unbraced frame subjected to a lateral load H. For the sake of the illustration, the columns are considered to have widely different axial loads. Column 1, with no axial load, provides lateral bracing, so does column 2, but somewhat less than it would have if its axial load P_2 had been zero. In most unbraced building frames, columns will typically be reasonably similar and with moment diagrams such as that for column 2. Column 3, with a high axial load, requires bracing from columns 1 and 2, in order for it not to fail in a sidesway mode. It "leans", in other words, on the others for its lateral stability. For column 4 (pinned at both ends), such bracing is absolutely necessary for it to remain laterally stable under even the smallest vertical loading. Therefore, column 4 represents the extreme case of a "leaning" column.

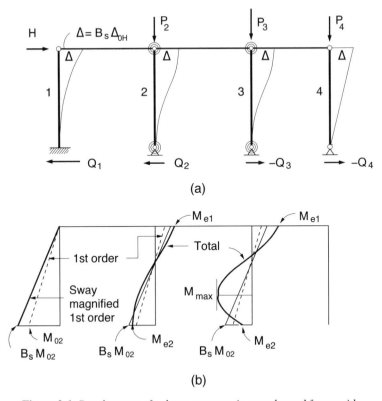

Figure 2.6. *Development of column moments in an unbraced frame with bracing and braced columns*

In other cases, column lengths may also be different (common in bridges across straits, etc). The frame could also have had an external bracing, without changing the general considerations made above. The development of maximum moments in such different columns has been discussed in the literature (Lai and MacGregor [LAI 83a], Hellesland [HEL 08b]). An overview in line with the latter reference is given below.

2.4.3.2. *Bracing column with no local second-order effects*

Due to the second-order effects, lateral sway deflection increases, from a first-order value due to H of Δ_{0H} to the total value:

$$\Delta = B_s \Delta_{0H} \qquad\qquad [2.8]$$

where B_s is a *sway magnification factor*. It reflects global second-order effects from all interconnected columns (on the same level, or story, in multilevel frames), and affects all columns (at the same level). Such sway magnifiers for multibay frames (or stories in multistory buildings) are presented in Chapter 3. Some of them are very accurate.

Moments in column 1, with no axial load, and therefore with no local second-order effects, increase proportionally with the global sway magnifier B_s, from M_0 to:

$$M_0^* = B_s M_0 \tag{2.9}$$

Similar moments are shown by dashed lines also for columns 2 and 3. They are global second-order *sway–modified first-order moments* [HEL 08b]. They include global second-order effects, but no local second-order effects.

2.4.3.3. *Bracing column with moderate local second-order effects*

Column 2, with end moments and moderate axial loads, has some local effects that affect the moments, but the maximum moment is still at a column end. The end moments, M_e, and the maximum moment M_{max} may be written as:

$$M_{ei} = B_{ei} M_{0i}^* = B_{ei} (B_s M_{0i}) \qquad (i = 1, 2) \tag{2.10}$$

$$M_{max} = B_m (B_s M_{02}) \quad \text{where} \quad B_m = B_{e2} \tag{2.11}$$

Here, B_{ei} is an end moment correction factor for *local* second-order effects. Simple B_{ei} expressions are available for some special cases [HEL 09a], but general, reasonable simple expressions are not available. However, the simplification $B_{ei} = 1$ is normally quite acceptable, and adopted in codes. Maximum moment in column 2 is at the end with the largest end moment, which per definition is denoted end 2.

For a single unbraced column, B_s is a function of the single column only. For such cases, it is possible to derive more accurate expressions for the product $B_m B_s$. This is done for some unbraced columns in section 2.5.1, and there denoted B_{m0}.

2.4.3.4. *Braced column with significant local second-order effects*

Column 3, with end moments and very high axial loads, experiences a lateral sway like that in the other columns, but can be considered braced at this sway. This is not an uncommon case, although it is more typical in fully braced frames. In the illustration, the column develops maximum moment between ends, similar to that shown for the fully braced column in Figure 2.5. The maximum moment can symbolically be expressed by:

$$\begin{aligned} M_{max} &= B_m \, M_{02}^* = B_m \, (B_s M_{02}) \\ &= B_{m1} \, C_m (B_s M_{02}) \end{aligned}$$

[2.12]

where $B_m = B_{m1} C_m$ is the local *moment magnification factor*, that reflects second-order moment effects in the individual column. In approximate methods, this factor can be expressed in terms of the column's braced buckling load (braced effective length factor). As will be shown in section 2.5.2, B_m can conveniently be expressed in terms of B_{m1}, which is the moment magnification factor for the column with a uniform moment distribution (indicated by the additional subscript 1), and C_m, which is a moment gradient factor that corrects for the difference between the uniform moment distribution and the real moment distribution.

Alternatively, the moment

$$M_{0e} = C_m M_{02}$$

[2.13]

can be considered an equivalent, uniform first-order moment that, when combined with B_{m1}, has the same effect on the maximum moment as the largest first-order end moment combined with B_m. M_{0e}^* is the corresponding equivalent sway–modified uniform first-order moment.

If the frame, in addition to the horizontal (sway) load, had been subjected to gravity (vertical) loads, first-order frame moments at column ends can, for instance, be obtained in the classical manner as the sum of those (M_{0b}) obtained for the fully braced (no-sway) frame subjected to all vertical and horizontal loads, and those (M_{0s}) obtained for the unbraced frame subjected to the horizontal bracing (holding) load from the braced case. Subscripts b and s have been adopted to denote the fully braced case and sway (unbraced) case, respectively.

The approximate maximum moment expression in equation [2.14] can then be extended to:

$$
\begin{aligned}
M_{max} &= B_m\, M_{02}^* \\
&= B_m\,(M_{0b} + B_s M_{0s})_2 \\
&= B_{m1}\, C_m (M_{0b} + B_s M_{0s})_2
\end{aligned}
\qquad [2.14]
$$

In practice, it may be inconvenient to analyze a structure by first considering it with a fictitious bracing, and next without this bracing. For structures with reasonable symmetry in geometry and vertical loading, there will be little sidesway. In such cases, M_{0b} above can, consequently, be taken as moments due to gravity (vertical) loads on the real structure. The M_{0s} moments are, as before, those due to horizontal loads (including inclination imperfections). If significant unsymmetry is present, the M_{0b} moments due to gravity loads on the unbraced structure, will include a sidesway contribution. The B_s factor can be adjusted to account for this. For additional discussion of these and similar formulations, see Hellesland [HEL 08b].

2.4.4. *Moment equilibrium at joints*

In unbraced frame structures, or frames with flexible lateral bracing, sidesway will be due to the wind loading and imperfection, and global effects due to the vertical gravity loading acting on these lateral deflections. The restraining, flexural elements (beams, floors, foundations) at column joints must equilibrate total column moments (first- and second-order) to maintain frame stability. It is, therefore, very important to consider second-order effects also in the design of restraining elements.

This is most important when the restraining flexural elements (beams, floors, foundations) at column joints are designed for moments that equilibrate the column moments. This will be the case for columns in unbraced frames, or frames with flexible lateral bracing, where the restraining beams, etc., must equilibrate the overturning moments due to the wind loading and imperfections (inclinations), and global effects due to the vertical gravity loading acting on these lateral deflections. Figure 2.6 illustrates such a case, and also the single column in Figure 2.4, where it is the foundation that must equilibrate the increased column end moment.

It is also the case for braced frames, in particular for beams restraining the exterior columns in a multi-story frame, or the columns in single-story frames, such as those illustrated in Figure 2.5. At the lower joint, the column moment increases due to second-order local effects, and the beam moment must, consequently, increase by the same amount. At the upper joint, with an externally applied joint moment, second-order effects cause the column moment to decrease. To balance the external moment, the beam moment must increase by the same amount.

If the beams are designed for the smaller first-order moments only, plastic hinges may form in the beams, near the joints (assuming the beams have sufficient rotation capacity). As a result, a mechanism may form, and collapse takes place, at a loading less than the required design loading.

For interior columns in multi-bay, multi-story braced frames, the situation is not as critical. Such columns are designed assuming the worst load pattern. Also, the beams are designed based on a moment envelope that generally gives beam moments at joints that exceed the moment corresponding to the most unfavorable moment loading for the column. Thus, for the most unfavorable column loading case, the beams will, in such cases, have a reserve strength. This beams reserve strength may be able to accommodate considerable changes in column end moments before it will develop plastic hinges and form a mechanism.

2.5. Maximum moment formation

2.5.1. *Maximum first- and second-order moment at the same section*

2.5.1.1. *Magnifier-based derivation*

In many cases, maximum first- and second-order moments will form at the same section. This will often be the case for laterally loaded individual columns, for columns in lateral loaded unbraced frames and transversely loaded beam columns. In such cases, it is easy to obtain reasonably accurate approximate expressions for maximum moment. Consider, for instance, the cantilever column in Figure 2.4 with a maximum first-order moment M_0 at the base and a corresponding first-order deflection Δ_0 at the top. The base may be clamped, as shown in the figure, or it be partially restrained.

It is well established that the total elastic lateral deflection Δ at the column top can be approximated by the first-order deflection Δ_0 times a sway deflection magnification factor B_s:

$$\Delta = B_s \Delta_0 \qquad\qquad [2.15a]$$

where

$$B_s = \frac{1}{1 - \alpha} \quad \text{and} \quad \alpha = \frac{N}{N_{cr}} \qquad\qquad [2.15b]$$

is the stability index, and N_{cr} the elastic critical load defined previously. From moment equilibrium, the total maximum moment at the base can be given by:

$$M_{max} = M_0 + M_2 = M_0 + \frac{N\Delta_0}{1 - \alpha} = \left(1 + \frac{N\Delta_0/M_0}{1 - \alpha} \right) M_0$$

or by

$$M_{max} = B_{m0} M_0 \qquad\qquad [2.16a]$$

where

$$B_{m0} = \left(1 + \frac{\beta}{(1/\alpha) - 1} \right) = \frac{1 + (\beta - 1)\alpha}{1 - \alpha} \qquad\qquad [2.16b]$$

and

$$\beta = \frac{N_{cr}\Delta_0}{M_0} = \frac{\pi^2 EI}{L_0^2} \cdot \frac{\Delta_0}{M_0} \qquad\qquad [2.16c]$$

Above, B_{m0} is a "basic" moment magnification factor that may be expressed by either of the two given forms.

2.5.1.2. *Curvature-based derivation*

The total top deflection is equal to the sum of the first-order Δ_0 and second-order Δ_2 deflections. For a constant EI and given moment distribution, the curvature distribution is known ($1/r = M/EI$), and the first- and second-order top deflections can be readily be calculated and expressed in a general form by:

$$\Delta_0 = \frac{L_0^2}{c_0} \cdot \frac{M_0}{EI} \quad \text{and} \quad \Delta_2 = \frac{L_0^2}{c_2} \cdot \frac{M_2}{EI} \qquad [2.17]$$

respectively, where c_0 depends on the curvature distribution from the first-order moment and c_2 on the distribution from the second-order moment. The second-order moment contribution can now be written as:

$$M_2 = N\Delta = N(\Delta_0 + \Delta_2) = \frac{NL_0^2}{c_2 \, EI}\left(M_0\frac{c_2}{c_0} + M_2\right)$$

or, when solving for M_2, as

$$M_2 = M_0 \frac{(c_2/c_0)\alpha}{1 - \alpha}$$

Thus, the total moment becomes:

$$M_{max} = M_0 + M_2 = \left(1 + \frac{c_2/c_0}{(1/\alpha) - 1}\right)M_0 = \frac{1 + ((c_2/c_0) - 1)\alpha}{1 - \alpha} \cdot M_0$$

$$[2.18]$$

The second-order curvature distribution, due to the second-order moment, can normally be approximated by a parabolic or sinusoidal distribution. In the latter case (adopted in EC2 and discussed by Westerberg [WES 04]), the well–known $c_2 = \pi^2$ results. This is also the value implied in the approximate sway magnifier B_s in equation [2.15].

By adopting $c_2 = \pi^2$ and substituting the first-order deflection Δ_0 from equation [2.17] into equation [2.16c], it is found that c_0 can be related to β through:

$$\beta = c_2/c_0 = \pi^2/c_0 \qquad [2.19]$$

With this relationship, it is seen that the maximum moment given by equation [2.18] becomes identical to the moment given before by equation [2.16].

2.5.1.3. β and c_0 factors

The β and c_0 factors are interrelated and dependent on the curvature distribution, which in the case with a constant EI has the same form as the moment distribution.

Typical β and corresponding c_0 values are summarized in Figure 2.7. These are computed with $L_0 = 2L$ for the cantilever column and $L_0 = L$ for the pinned-end column.

As an example, equation [2.16c] is applied to the simply supported member with uniform loading (Case g, Figure 2.7). For this case, the well–known values of $\Delta_0 = 5qL^4/(384EI)$, $M_0 = qL^2/8$ and $L_0 = L$ result. With these values, equation [2.16c] gives $\beta = 1.028$, and equation [2.19] gives $c_0 = 9.6$. These are the same values given in the figure.

For other cases with other boundary conditions, care must be exercised. For instance, for the braced, clamped-ends case (Case d) in Figure 2.7, with $L_0 = 0.5L$, Δ_0 should be introduced with a deflection value measured from the line through the inflection points (a distance $L/2$ apart). Similarly, for the unbraced, clamped-ends case (Case e) with $L_0 = L$, Δ_0 should be introduced with a deflection value measured from the horizontal line through the inflection point at mid–span. With these values (half the total values), equation [2.16c] gives the correct result.

Because Cases a through to f can be broken down into cantilever columns with triangular moment diagrams, it was to be expected that they all have the same β and c_0 values.

For the uniform first-order moment case in Figure 2.7 (Case a), $\beta = 1.23$ is found using equation [2.16c]. This case is also considered in an exact theory context later by equation [2.31] ($M = M_0/\cos(pL/2)$), which indicates that $\beta = 1.25$ would have been a better value for the upper range of α values that may be encountered in practice. In general, the accuracy of these approximations are normally within 2%.

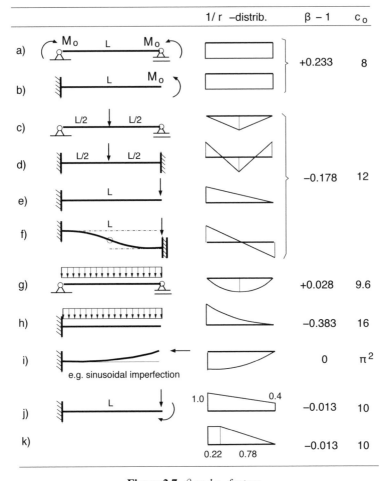

Figure 2.7. β and c_0 factors

2.5.2. Maximum first- and second-order moment at different sections

2.5.2.1. Maximum moment derivation

It is of interest to establish the theoretical basis for the maximum moment formulation for columns with unequal end moments and without transverse loads between ends. In such cases, maximum first-order and maximum second-order moments will generally result at different sections.

Consider an initially straight column, with constant bending stiffness EI in the plane of bending, and subjected to end moments M_{tA} and M_{tB}, shear force V and axial force N as shown in Figure 2.8. The subscript t is added for clarity to end moments to indicate that they are the total moments (first-order plus second-order effects) obtained from second-order theory. The moment distribution along the column can readily be obtained from the differential equation of the elastic curve, and maximum moments can conveniently be determined and expressed in terms of the end moments.

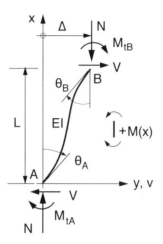

Figure 2.8. *Column definition and sign convention*

Assuming negligible shear deformations, the well–known second-order differential equation is given by:

$$v(x)'' = -M(x)/EI \qquad [2.20]$$

The displacement v and moment M at an arbitrary section x can be determined from this equation and expressed by:

$$v(x) = C_1 \sin px + C_2 \cos px - (M_{tA} + Vx)/N \qquad [2.21]$$

$$M(x) = -EIv(x)'' = p^2 C_1 \sin px + p^2 C_2 \cos px \qquad [2.22]$$

where

$$p = \sqrt{(N/EI)} \qquad [2.23]$$

The constants can be determined from the boundary conditions. For the moment expression, which is the one of interest here, the end moment conditions $M(0) = M_{tA}$ and $M(L) = -M_{tB}$, give $p^2 C_2 = M_{tA}$ and $p^2 C_1 = -(M_{tB} + M_{tA} \cos pL)/\sin pL$. With these, and choosing to express $M(x)$ in terms of the *end moment M_{tA} at the origin*, equation [2.22] becomes:

$$M(x) = M_{tA} \left[\frac{\mu_t - \cos pL}{\sin pL} \sin px + \cos px \right] \qquad [2.24]$$

where

$$\mu_t = -M_{tB}/M_{tA} \qquad [2.25]$$

is defined as the negative ratio between the end moments. Note that this definition implies that μ_t becomes positive when the end moments act in opposite directions (clockwise and counterclockwise) and, thus, inflicts curvature to the same side of the column at its ends.

The location(s) x_m of the peak moment(s) along the trigonometric moment curve can be found from the peak condition, $dM(x)/dx = 0$. From this, the peak location(s) can be expressed by:

$$\tan px_m = \frac{\mu_t - \cos pL}{\sin pL} \qquad [2.26]$$

with solutions $px_m = px_{m,0} + n\pi$, where $n = 1, 2, ...$

The peak moment can be expressed in either of the following two forms:

$$M_{peak} = \frac{M_{tA}}{\cos px_m} = \pm \left| M_{tA} \frac{\sqrt{1 + \mu_t^2 - 2\mu_t \cos pL}}{\sin pL} \right| \qquad [2.27]$$

The first form above, obtained directly by substituting equation [2.26] into equation [2.24], gives correct peak moment signs. The second formulation may not do so.

The second formulation can be obtained in a well–known manner (e.g. Galambos [GAL 68]) by letting the numerator and denominator of equation [2.26] represent the opposite and adjacent short sides, respectively, in a right–angled triangle, and then expressing $\cos px_m$ by the adjacent side and the hypotenuse (Pythagorean theorem). For restrained columns, pL may become greater than π and $\sin pL$ become negative. This gives a negative magnifier. The absolute signs on the magnified moment expression are included for this reason. The sign is generally of minor interest for the typical cases of symmetrically reinforced columns.

The peak moment, equation [2.27], may be located between column ends, in which case the maximum moment becomes equal to the peak moment. If the peak moment is located on the trigonometric moment curve outside the column length, the maximum moment becomes equal to the larger of the two end moments. These cases can be determined as follows:

1) If $px_{m,0} \leq 0$ and $px_{m,0} + \pi \geq pL$, the maximum moment is at an end, and given by $M_{max} = \pm \max(|M_{tA}|, |M_{tB}|)$.

2) If $0 \leq px_{m,0} < pL$, or if $px_{m,0} < 0$ and $px_{m,1} = px_{m,0} + \pi < pL$, the maximum member moment will be between ends and given by $M_{max} = M_{peak}$. This will always be so when i) $pL > \pi$ (i.e. $N > N_E$), or ii) if $pL < \pi$ and $\mu_t > \cos pL$.

Let us now, for simplicity, define M_{tA} to be the one with the larger absolute value of the two total end moments. Then, the maximum moment for the considered column (without lateral loading between ends), can be expressed by

$$M_{max} = B_{t\,max} M_{tA} \qquad\qquad [2.28]$$

where

$B_{t\,max} = 1$ when the maximum moment is at end A, and otherwise, $B_{t\,max} = B_{t\,peak}$ (equation [2.27]).

The peak moment factor $B_{t\,peak}$ can be rewritten as a product of two terms:

$$B_{t\,peak} = C_m B_{t\,peak,1} \qquad\qquad [2.29]$$

where

$$C_m = \sqrt{\frac{1 + \mu_t^2 - 2\mu_t \cos pL}{2 - 2\cos pL}} \qquad [2.30]$$

and

$$B_{t\,peak,1} = \frac{\sqrt{2 - 2\cos pL}}{|\sin pL|} = \frac{1}{\cos(pL/2)} \qquad [2.31]$$

The first term (C_m) becomes equal to 1.0 for $\mu_t = 1$, that is in the case when the two end moments are equal and acting in opposite directions (providing tension on the same side of the column at its ends). The second term ($B_{t\,peak,1}$) is, in other words, the moment magnification factor for a column with equal and opposite end moments. The subscript 1 is included to reflect $\mu_t = 1$.

C_m is often referred to as an equivalent, uniform moment factor, or a moment gradient modification factor.

C_m as defined by equation [2.30] is presented in Figure 2.9 for selected axial force levels $\alpha_E = N/N_E$ less than 1.0, as function of the end moment ratio μ_t. Also shown is the locus through points with the same total moment magnification as given by equation [2.29] for $B_{t\,peak} = 1$, 1.1, 1.5 and ∞. For $B_{t\,peak} = 1$, the maximum moment is just at the column end.

For design purposes, Massonnet [MAS 59] proposed a simplified expression defined by:

$$C_m = \sqrt{0.3\mu_t^2 + 0.4\mu_t + 0.3} \qquad [2.32]$$

for the cases illustrated in Figure 2.9. A still simpler approximation was given by Austin [AUS 61]:

$$C_m = 0.6 + 0.4\mu_t \geq 0.4 \qquad [2.33]$$

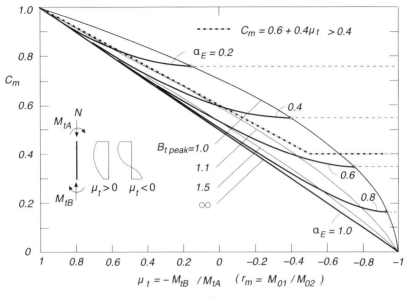

Figure 2.9. C_m *curves*

An excellent approximation of equation [2.31] for $N < N_E$ ($pl < \pi$) is given by:

$$B_{t\,peak,1} = \frac{1 + 0.25\alpha_E}{1 - \alpha_E} \qquad [2.34]$$

This is a common approximation for a pinned-end column (see, for instance, Lai *et al.* [LAI 83b] and Galambos [GAL 68]).

Then, with this approximation and Austin's C_m expression, a $B_{t\,max}$ factor that covers cases with maximum moment either at or between ends, can for $\alpha_E < 1.0$ be approximated by

$$\left.\begin{aligned}
M_{max} &= B_{t\,max}M_{tA} \\
B_{t\,max} &= C_m \cdot \frac{1 + 0.25\alpha_E}{1 - \alpha_E} \geq 1.0 \\
C_m &= 0.6 + 0.4\mu_t \geq 0.4 \\
\mu_t &= -M_{tB}/M_{tA} \\
\alpha_E &= N/N_E \qquad (N_E = \pi^2 EI/L^2)
\end{aligned}\right\} \qquad [2.35]$$

It is recalled that M_{tA} is taken as the largest of the two end moments. The lower limitation (≥ 1.0) on $B_{t\,max}$ is required in order to also cover cases when the maximum moment is at a column end. It will be discussed further below for the pinned-end column.

The theory and approximations presented above are valid for columns with any lateral and rotational support conditions. The columns can be fully braced, or partly braced with some sway.

2.5.2.2. *Application to pinned-end columns*

For the special case of columns pinned at both ends, the end moments are not affected by second-order effects. Therefore, total end moments are identical to the applied (first-order) moments. Figure 2.9 and the approximation given by equation [2.35], valid for $\alpha_E < 1.0$, are consequently also applicable to pinned-end columns with end moments $M_{A(B)}$.

Typical results for a pinned-end column are given in Figure 2.10. End moment ratios of $\mu_t = +0.8$, 0, -0.8, as indicated by the inserts above the curves in the figure, are considered. Exact results, computed by equation [2.28], are shown by the full lines. Initially, the maximum moment is at end A, as reflected by $B_{t\,max} = 1$. At some load level α_E, maximum moment develops away from the end. This is reflected by $B_{t\,max}$ values increasing beyond 1.0. For an initially uniform moment distribution, $\mu_t = 1.0$ (not shown in the figure), the maximum moment (at mid-height) starts increasing beyond 1.0 immediately as α_E increases from zero. A moment gradient delays the development of maximum moment away from the end. As seen in the figure, an increasing moment gradient requires an increasing α_E value before this happens.

The approximation given by $B_{t\,max}$ in equation [2.35], but without the lower limitation, is shown by the dashed (broken) lines. It is quite clear from these results that it is necessary to require $B_{t\,max}$ to become greater than or equal to 1.0. Without this restriction, significant underestimation of maximum moments may result.

The approximate results are very good for single curvature cases with nearly uniform initial moment distributions (C_m close to 1.0). They become increasingly conservative, compared to exact results, with increasing moment gradient and initial curvature distributions approaching perfect antisymmetric bending ($C_m = -1.0$). For the three cases with $\mu_t = +0.8,\ 0,\ -0.8$ in Figure 2.10, C_m becomes 0.92, 0.6 and 0.4, respectively. In the latter case, the lower limit of 0.4 on C_m applies. It is clearly very conservative, as can be seen also in Figure 2.9. According to the figure, a correct C_m value should be less than 0.2 at $\mu_t = -0.8$. The approximate C_m in equation [2.35] without the 0.4 limitation gives $C_m = 0.28$. Use of these values would have given much better correspondence between the approximate and exact results for the $\mu_t = -0.8$ case.

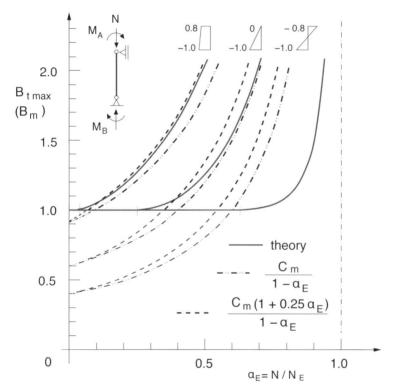

Figure 2.10. *Moment magnification factors for pinned-end column with three different moment gradients: Comparison between exact ("theory") and approximate factors*

An approximation of the moment magnifier for the uniform first-order moment case in equation [2.34], can also be obtained from equation [2.16b] for a pinned-end column. With $\beta - 1 = 0.23$ (Figure 2.7), equation [2.16b] gives

$$B_{m1} = \frac{1 + 0.23\alpha_E}{1 - \alpha_E} \qquad [2.36]$$

which is identical to equation [2.34] except for the numeral 0.23 instead of the more accurate numeral 0.25. EC2 has adopted 0.23 in its nominal stiffness methods.

Also shown in the figure are results obtained with the simpler approximation given by

$$B_{t\,max} = C_m \cdot \frac{1}{1 - \alpha_E} \qquad [2.37]$$

This is the most common approximation, it is believed, and allowed by EC2. It is adopted by the ACI 318 code, with the C_m approximation given in equation [2.35], but without the lower 0.4 limitation.

It should be noted that the total end moment ratio μ_t and the $B_{t\,max}$ factor for a pinned-end column are identical to the first-order moment ratio r_m and the B_m factor defined below, respectively.

2.5.2.3. Adaption to end moments from first-order theory

An extension of the pinned-end column results can be achieved in an approximate manner in accordance with the so-called "effective length method", proposed by Winter [WIN 54] based on work by Lee [LEE 49] and Bijlaard et al. [BIJ 53]. A rigorous assessment of the method is given by Lai et al. [LAI 83-b].

This method represents an extension of the pinned-end column approach. It involves: (1) the replacement of the applied end moments by end moments obtained from a first-order analysis, and (2) accounting for rotational restraints at column ends by replacing the column length L by the effective length L_0 of the column considered braced.

The first-order end moments have previously, and in line with EC2 notation, been denoted M_{01} and M_{02}. They are defined as positive when they act in opposite directions at the two ends, that is, one of these moments acts in the opposite direction to that defined as positive in Figure 2.8. With these end moments, the end moment ratio μ_t above should be replaced by:

$$r_m = M_{01}/M_{02}$$

(without a minus sign, in line with the sign convention for this moment ratio and the EC2 convention). A positive ratio implies single curvature first-order bending.

The resulting magnifier has traditionally been considered valid for braced columns only. However, as discussed by Hellesland [HEL 08b], and in conjunction with the frame with sway in Figure 2.6 (column 3), it can be extended to any column without transverse loading between ends, by replacing M_{01} and M_{02} by the (global second-order) sway–modified first-order end moment M_{01}^* and M_{02}^*.

Thus, in approximate theory for framed columns, the moment-magnifier approach can then in the general, framed column case, be defined by:

$$\left.\begin{aligned}
M_{max} &= B_m M_{02}^* \\
B_m &= C_m \cdot \frac{1 + 0.25\alpha}{1 - \alpha} \geq 1 \\
C_m &= 0.6 + 0.4\, r_m \geq 0.4 \\
r_m &= M_{01}^*/M_{02}^* \\
\alpha &= N/N_{cr} \qquad (N_{cr} = \pi^2 EI/L_0^2)
\end{aligned}\right\} \qquad [2.38]$$

Here, L_0 is the effective length factor of the column considered fully braced, N_{cr} is the corresponding critical (buckling) load, B_m is the moment magnifier to be applied to M_{02}^*, which is the larger of the two second-order sway-modified first-order end moments. This, and the smaller one, M_{01}^*, are discussed previously (section 2.4.3) and defined by

$$\begin{aligned}
M_{02}^* &= (M_{0b} + B_s M_{0s})_2 \\
M_{01}^* &= (M_{0b} + B_s M_{0s})_1
\end{aligned} \qquad [2.39]$$

By comparing with equation [2.14], B_m above corresponds to $B_{m1} C_m$ in that equation.

The simpler B_m expression given by:

$$B_m = C_m \cdot \frac{1}{1 - \alpha} \geq 1 \tag{2.40}$$

will generally be acceptable. It is, as mentioned above, the more common approximation in design codes, and it is also accepted by EC2.

Exact end moments and the maximum moment, given dimensionless in terms of the largest first-order end moment M_{02}, are shown versus the load index α_E in Figure 2.11 (full lines) for the restrained, braced column shown

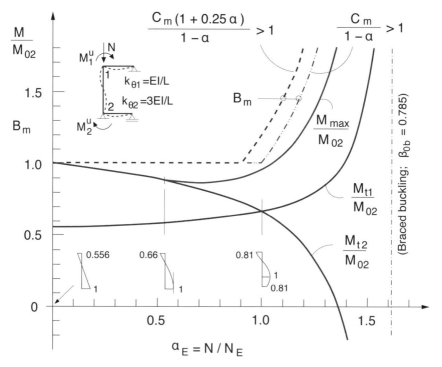

Figure 2.11. *Variation of maximum and end moments versus axial load in a restrained column*

by the insert in the figure ($M_{02}^* = M_{02}$ because $B_s = 1$). The exact, braced effective length factor of the column is $\beta_{0b} = 0.785$. Thus, braced buckling is obtained at $\alpha_E = 1/\beta_{0b}^2 = 1.622$. The moments $M_{1(2)}^u$ are unbalanced moments applied at the joints and provide a first-order end moment ratio $r_m = M_{01}/M_{02} = -0.556$ and a first-order moment distribution along the column as shown in the figure.

Because of the rotational end restraints, the end moments change with increasing load. At end 1, the moment increases whereas at end 2, with the stiffer restraint and with the larger first-order end moment, the moment decreases (and eventually reverse direction). The maximum moment is at end 2 until at some point, maximum moment forms away from the end. The moment distribution corresponding to this point is illustrated schematically, with relative moment values, in the figure.

Also included in the figure are approximate maximum moment predictions by equation [2.38] with $C_m = 0.4$ ($C_m = 0.60 + 0.4r_m$ becomes 0.38). Because there is no sway, $M_0^* = M_0$ in this case. For the purpose of plotting the approximate solution, the stability index α in equation [2.16b] can be expressed in terms of α_E by $\alpha = \alpha_E \beta_{0b}^2 = 0.616\alpha_E$.

It can be seen that the approximate prediction is quite conservative. Results obtained with the simpler B_m expression, equation [2.40], also shown in the figure, are conservative, though less so than equation [2.38]. However, the simpler B_m will normally lead to somewhat unconservative predictions for single curvature bending cases, similar to that seen for the pinned-end column in Figure 2.10.

Without the limitation "≥ 1" on B_m in equations [2.38] and [2.40], predictions would become grossly unsafe for intermediate and lower axial forces (B_m approaches $C_m = 0.4$ for axial loads approaching zero).

2.5.2.4. *The lower limit on the C_m approximation*

The EC2 code has adopted the C_m approximation above with 0.4 as a lower limitation. The same approximation was earlier adopted by the American ACI 318 code. However, in newer editions, including ACI 318-2008 and ACI 318-2011, the approximation:

$$C_m = 0.6 + 0.4\,r_m \qquad\qquad\qquad [2.41]$$

without any 0.4 limit, was adopted. This approximation terminates at a value of $C_m = 0.2$ in the perfect double curvature case ($r_m = -1$). From Figure 2.9, valid for $N < N_E$, it would seem that this is sufficiently conservative at least for pinned-end columns.

The choice of the 0.4 limitation in EC2 is not justified in the code. It may be argued, in the general case for restrained columns, that some additional conservativeness at high moment gradients may be considered appropriate to account for uncertainties in the assessments of end moment ratios, and possible consequences of sudden changes in the deflected shape at high moment gradients, when the central part of the column may "snap out" into a first buckling-like mode. For an elastic pinned-end column in perfect double curvature, this happens at $N = N_E$. Such phenomena, often called unwrapping or unwinding, are discussed in the literature, by Lai, MacGregor and Hellesland [LAI 83b], Hellesland [HEL 02a], and others. The question of whether equation [2.41] is sufficiently conservative to cover such problems should probably be considered in a future revision of the code.

2.5.3. *Curvature-based maximum moment expression*

2.5.3.1. *Maximum moment derivation*

A curvature-based approach in line with the major steps of the "Sinusoidal total eccentricity method" presented by Robinson *et al.* [ROB 75], is considered here. It provides a more fundamental basis than what is usual in the presentation of the simplified "nominal curvature" method, earlier referred to as the "model column method" (CEB-FIP Buckling Manual [COM 77]; CEB-FIP Model Code 1990 [COM 93]). The method is also described, with some additional information, in Mari and Hellesland [MAR 03].

Consider the rotational unrestrained (pinned-end) column with unequal end eccentricities e_{01} and e_{02} in Figure 2.12. By choosing the origin of the coordinate system at the section with maximum total load eccentricity e_{max} (first- plus second-order), the total eccentricity e at an arbitrary section x becomes:

$$e(x) = e_0(x) + w(x) = e_{max} \cos \frac{\pi x}{a} \qquad [2.42]$$

where $e_0(x)$ is the first-order eccentricity and $w(x)$ the deflection at the same section from the chord through the end eccentricity extremities. Then,

$$e_{01} = e(x_1) = e_{max} \cos \frac{\pi(-x_1)}{a} \quad ; \quad e_{02} = e(x_2) = e_{max} \cos \frac{\pi x_2}{a}$$

and

$$x_1 = -\frac{a}{\pi} \arccos \frac{e_{01}}{e_{max}} \quad \text{and} \quad x_2 = \frac{a}{\pi} \arccos \frac{e_{02}}{e_{max}}$$

Substitution into $x_2 - x_1 = L$ and solving for the half wavelength a yields:

$$a = \frac{\pi L}{\arccos \dfrac{e_{02}}{e_{max}} + \arccos \dfrac{e_{01}}{e_{max}}} \qquad [2.43]$$

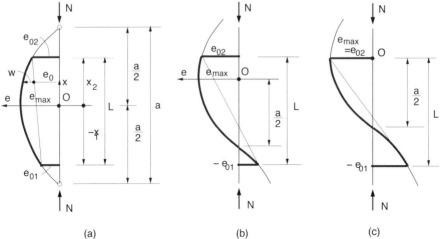

Figure 2.12. *Sinusoidal total eccentricity*

The curvature $1/r$ corresponding to e_{max} can now be obtained from the double derivative of equation [2.42] and expressed by:

$$\frac{1}{r} = -e''(0) = e_{max} \frac{\pi^2}{a^2} = \frac{e_{max}}{L^2} \left(\arccos \frac{e_{02}}{e_{max}} + \arccos \frac{e_{01}}{e_{max}} \right)^2 \quad [2.44]$$

This curvature expression has physical relevance only provided at the peak, e_{max}, is located within the column length. If it is located outside the column length, the maximum eccentricity is equal to the larger end eccentricity e_{02}. A case when these two coincide is shown in Figure 2.12(c). At this limit, the corresponding maximum curvature, $1/r_2$, is obtained by substituting e_{02} for e_{max} in equation [2.44]. This gives:

$$\frac{1}{r_2} = \frac{e_{02}}{L^2}\left(\arccos\frac{e_{01}}{e_{02}}\right)^2 \qquad [2.45]$$

Mari and Hellesland [MAR 03] suggested a parabolic approximation of equation [2.45] given by:

$$\frac{1}{r_2} = \frac{\pi^2 e_{02}}{4L^2}\left(\frac{e_{01}}{e_{02}} - 1\right)^2 \qquad [2.46]$$

An overall better approximation than the parabolic approximation above is found here to be given by:

$$\frac{1}{r_2} = \frac{\pi^2 e_{02}}{L^2}\left(1 - \sqrt{\frac{1 + e_{01}/e_{02}}{2}}\right) \qquad [2.47]$$

When the curvature increases beyond the $1/r_2$ limit, the load eccentricity increases beyond e_{02} as given by equation [2.44], which approaches asymptotically a slope L/π^2. Assuming simplified that e_{max} follows a response with this slope already from the curvature limit $1/r_2$, e_{max} can be written as:

$$e_{max} = e_{02} + \frac{L^2}{\pi^2}\left(\frac{1}{r} - \frac{1}{r_2}\right) \qquad [2.48]$$

This expression can be rewritten, in a form that facilitates easy comparison with the previous magnifier formulation, as:

$$e_{max} = C_{me}\,e_{02} + \frac{L^2}{\pi^2}\cdot\frac{1}{r} \qquad [2.49]$$

With the parabolic approximation in equation [2.46], C_{me} in equation [2.49] becomes:

$$C_{me} = 1 - \frac{1}{4}\left(\frac{e_{01}}{e_{02}} - 1\right)^2 \qquad [2.50]$$

and with the alternative approximation in equation [2.47]:

$$C_{me} = \sqrt{\frac{1 + e_{01}/e_{02}}{2}} \qquad [2.51]$$

Without showing the transition from an approximate $1/r_2$ assumption, Robinson *et al.* [ROB 75] proposed:

$$C_{me} = 1.45 - 0.05\left(4 - \frac{e_{01}}{e_{02}}\right)^2 \qquad [2.52]$$

To compare with the approximation of the differential equation based expression in equation [2.35], equation [2.49] can be written on magnification factor form. Multiplication with N, substitution of $1/r = M_{max}/EI$ and solving for M_{max} gives:

$$M_{max} = \frac{C_{me}}{1 - \alpha_E} M_{02} \qquad [2.53]$$

From comparison with equation [2.35], it is clear that it is $C_{me}/(1 + 0.25\alpha_E)$ that is comparable to C_m in equation [2.35] and in Figure 2.9.

Of the approximations, equation [2.51] seems to be the most accurate. For small load levels (α_E), corresponding to small e_{max}/e_{02} ratios, the approximation $C_{me}/(1 + 0.25\alpha_E)$ generally gives larger values than the exact C_m (equation [2.30]). For increasingly uniform bending, and increasing α_E values, the approximations may become unconservative, but not overly so for practical α_E values.

An improvement can be obtained if the asymptotic slope L^2/π^2 is replaced by L^2/c, where c should vary from about 8 to π^2 to give a more gradual change in slope. In principle, c should account for the term $1 + 0.25\alpha$ in the magnifier approach (reflects the difference between a column with a uniform first-order moment diagram and a sinusoidal first-order moment diagram).

2.5.3.2. Adaption to end moments from first-order theory

As for the magnifier approach, the adaption to columns with end restraints, the length L is replaced by the effective length L_0. The maximum moment expression can then be written as:

$$M_{max} = C_m M_{02} + \frac{L_0^2}{c} \cdot \frac{1}{r} \qquad [2.54]$$

This is the form adopted by EC2. The c factor is dependent on the curvature distribution from the total moment. It is different from the c_0 factor presented before (Figure 2.7), as c_0 reflected the effect of elastic (curvature) distribution due to the first-order moment only.

In principle, c should account for the term $1 + 0.25\alpha_E$ in the magnifier approach (reflecting the difference between a column with a uniform first-order moment diagram and a sinusoidal first-order moment diagram).

Just to get a closer grip on the similarities between the present and the magnifier approach, equation [2.54] can be rewritten as:

$$M_{max} = \frac{C_{m1}}{1 - \alpha \cdot (\pi^2/c)} M_{02} \qquad [2.55]$$

In order that this moment should give the same results as the differential equation-based approximation, equation [2.38], then the relationship:

$$\frac{1 + 0.25\alpha}{1 - \alpha} = \frac{1}{1 - \alpha \cdot (\pi^2/c)}$$

must be satisfied. Solving for c gives

$$c = \frac{1 + 0.25\,\alpha}{1.25} \pi^2 \qquad [2.56]$$

For α = 0, 0.2, 0.4, 0.6 and 1.0, c values of 8.0, 8.3, 8.7, 9.1 and π^2 are obtained.

In terms of the curvature-based formulation, equation [2.54], the second term with c = 8 corresponds to the maximum deflection for a uniform curvature distribution. Such a distribution is associated with small second-order effects, which is just the case with α_E values close to zero. With increasing second-order contributions (increasing α_E values), the curvature distribution deviates to an increasing extent from the uniform case, and the c value will increase.

It should be noted that the curvature-based formulation allows for the use of c values for real cases, that may greatly exceed those predicted by equation [2.56]. This is a strength of this formulation as compared to the elastically based moment magnifier formulation.

Recognizing the difficulty of determining a "correct" c value, EC2, and CEB-FIP model codes and other codes before, accepted the use of $c = 10 \, (\approx \pi^2)$. That was the value normally used in practical design work.

Furthermore, EC2 has adopted the Austin approximation for C_m (equation [2.33]) also in the simplified nominal curvature method. This and the $c = 10$ assumption are equivalent to setting $1 + 0.25\alpha$ in equation [2.38] equal to unity.

2.5.4. *Unbraced frame application example*

An example demonstrating some aspects of the application of the multiplier method to unbraced frames, as discussed in section 2.4.3 (in conjunction with equation [2.14]) and section 2.5.2 (in conjunction with equation [2.38], may be useful. For this purpose, the simple case of a single cantilever column, fully fixed at the base and pinned at the top, is considered. One reason for this choice, apart from being simple, is that the maximum moment for this column can also be calculated quite accurately using equation [2.16] and used for comparison.

Case I: The column is defined by the insert (I) in Figure 2.13. It is subjected to an applied clockwise end moment M_a at the top.

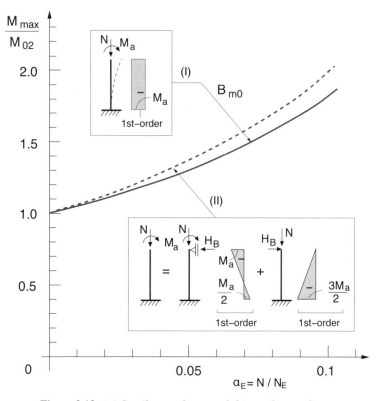

Figure 2.13. *(a) Cantilever column, and (b) cantilever column treated as an unbraced "frame"*

The maximum moment is given by equation [2.16] for $\beta = 1.23$ (Figure 2.7), $L_0 = 2L$ and $\alpha = \alpha_s = N/N_{cr,s} = 4\alpha_E$ for the unbraced cantilever. This gives, with $M_{02} = -M_a$,

$$B_{m0} = \frac{M_{max,I}}{M_{02}} = \frac{1 + 0.23 \cdot 4\alpha_E}{1 - 4\alpha_E} \qquad [2.57]$$

Case II: Following the same manner an unbraced frame would be treated by approximate methods, defined by equation [2.38], the calculation is carried out in two steps as indicated by insert (II) in Figure 2.13.

The system is first considered braced, giving first-order moments $M_{01,b} = -M_a$, $M_{02,b} = +0.5M_a$ and first-order bracing force $H_B = 3M_a/2L$. By applying H_B in the opposite direction to the unbraced system, the first-order sway moments $M_{01,s} = 0$ and $M_{02,s} = -1.5M_a$ are obtained.

Sway-modified end moments (equation [2.39]) then become:

$$M_{01}^* = M_{01,b} + B_s M_{01,s} = -M_a$$
$$M_{02}^* = M_{02,b} + B_s M_{02,s} = +0.5M_a + B_s \cdot (-1.5M_a) = (0.5 - 1.5B_s)M_a$$

where $|M_{02}^*| > |M_{01}^*|$ (according to the numbering convention).

For the effective length $L_0 = 2L$, the sway magnifier becomes:

$$B_s = 1/(1 - \alpha_s) = 1/(1 - 4\alpha_E)$$

The moment gradient correction factor, defined by:

$$C_m = 0.6 + 0.4\frac{M_{01}^*}{M_{02}^*} = 0.6 + 0.4\frac{-1}{0.5 - 1.5\,B_s} \geq 0.4$$

will vary between 1.0 and 0.83 for B_s values between 1 and 1.5.

The maximum moment magnifier B_m, calculated with braced effective length $L_0 = 0.7L$, giving $\alpha = \alpha_b = N/N_{cr,b} = 0.49\alpha_E$, becomes:

$$B_m = C_m \frac{1 + 0.23 \cdot 0.49\alpha_E}{1 - 0.49\alpha_E} \geq 1.0$$

where 0.23 instead of 0.25 is used in the nominator for the sake of consistency with equation [2.57]. This B_m magnifier varies between 1.0 and 0.85 for α_E between 0 and 0.1 (for the latter value, $B_s > 1.5$). Thus, since the maximum moment magnifier is less than 1.0, the lower limit of $B_m = 1.0$ applies. This is consistent with the fact that the maximum moment is at the end.

Then, with $M_{02} = M_{02,b} + M_{02,s} = +0.5M_a - 1.5M_a = -M_a$,

$$\frac{M_{max,II}}{M_{02}} = B_m M_{02}^* = 1.0 \cdot (0.5 - 1.5 \cdot B_s) = \frac{1 + 2\alpha_E}{1 - 4\alpha_E} \qquad [2.58]$$

This ratio is plotted versus α_E values of practical interest in Figure 2.13 (broken line). Compared to the results of equation [2.57] (full line), which are close to exact results, equation [2.58] results are about 6% greater at $\alpha_E = 0.06$ ($B_s \approx 1.5$).

The reason for this discrepancy is mainly due to errors originating in the braced part of the calculation (deviation between first- and second-order end moment and bracing force). With correct (second-order theory) bracing force H_B, the sway part of the calculation would have been very accurate.

In the example, the applied moment (from "gravity loading") was the only cause for sidesway. In multi-bay unbraced frames, with reasonably symmetrical gravity loading, the bracing force will be small, and so will corresponding errors due to inaccuracies in the bracing force assessment.

2.6. Local and global slenderness limits

2.6.1. *Local, lower slenderness limits – general*

2.6.1.1. *Background*

In a large number of cases, second-order load effects are small and can be neglected in regular building structures. Traditionally, most reinforced concrete codes have included slenderness limits to indicate when second-order effects can be neglected in individual compression members. So also in EC2, which states that "second order effects can be ignored when they are less than 10% of the corresponding first-order effects". Other codes have 5% as a criterion for such neglect. An overview of such limits, here denoted "lower slenderness limits", is given in Hellesland [HEL 05a] for a number of national and international codes.

In earlier codes, such limits were often given in terms of geometric slenderness, L/h or L_0/h, and mechanical slenderness, $\lambda = L_0/i$. Later, the moment gradient, expressed through the first-order end moment ratio, was included. These parameters do not give a very representative expression for a reinforced concrete column's slenderness, because they neglect the relatively strong effect of the normal force and the reinforcement on the sectional stiffness.

In recognition of this, the Norwegian Standard NS 3473:1989 [NOR 89] adopted, as the first code, limits that also included axial force and reinforcement in addition to mechanical slenderness and end moment ratio. A limit for columns with sustained (long-term) loads was also given. The limits were proposed by Hellesland [HEL 90a] based on a criterion related to a 5%

reduction in moment capacity. Later, less conservative limits based on the same principles were proposed [HEL 02a, HEL 02b, HEL 05a]. An update of these to a 10% criterion was adopted in the National Annex (NA.5.8.3.1) of the Norwegian adaption of the EC2 [EUR 08b]. They will, for simplicity, be referred to as the "NS-EC2" limits.

It is believed that these slenderness limit formulations have had significant impact on the so-called "recommended" slenderness limits in EC2:2004 (for countries that do not have their own rules). The EC2 rules are a function of creep in addition to the other parameters mentioned above. Both these and the NS-EC2 limits will be considered below.

Still another set of significantly different limits were developed by Mari and Hellesland [MAR 03, MAR 05b] from a curvature-based formulation. They include all the parameters of the limits above, provide very good assessments and can easily be adjusted to various criteria (discussed below).

2.6.1.2. *Criteria*

In structures with nonlinear material properties, an objective lower slenderness criterion can most appropriately be related to an acceptable percentage reduction in load carrying capacity. Within such a general definition, it is possible to envisage several more detailed criteria. These include [HEL 02a, HEL 02b, HEL 05a, HEL 08] a specified percentage reduction:

(a) in moment capacity for an applied constant axial load;

(b) in axial load capacity (and moment capacity) for an applied constant axial load eccentricity;

(c) in axial load capacity for an applied constant moment.

These three different criteria, in terms of capacity reductions, are illustrated schematically in Figure 2.14 (from [HEL 02b]). The curves labeled a, b and c in this axial load–moment $(N - M)$ interaction diagram, correspond to the 10% capacity reductions defined above.

Slender member strengths based on these three criteria, and corresponding slenderness results, may become very different. This is especially so at higher axial load levels where criteria b and c are seen to be considerably more

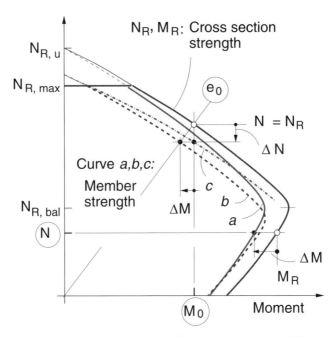

Figure 2.14. *Capacity reductions (10%) according to three different slenderness limit criteria*

generous (allowing greater slenderness limits) than criterion a. Trends in approximate slenderness limit predictions according to the three criteria are shown in Figure 2.15. The corresponding load applications are shown by the inserts. M_c is the design column moment taken equal to the ultimate design moment capacity M_{Rd}.

It can be seen that the different criteria may give widely different limits. Criterion b is the most liberal (in the sense that it provides the highest slenderness limits) at higher axial load levels. It is slightly more conservative (i.e. provides smaller slenderness limits) than criterion a at lower load levels (below the balanced load at about $\nu = 0.43$). Criterion c gives results mostly between the two others. The lower terminations of the criterion c curves correspond to an applied moment approximately equal to the pure moment capacity. As defined here, criterion c is not relevant for larger moments.

With increasing moment gradient (decreasing M_{01}/M_{02} ratio), the criterion b and c curves approach the horizontal line corresponding to criterion a. This points to the use of criterion a as a single, acceptable criterion, for the general case, when loading type is not known. Possible effects of creep, that are not considered explicitly above, are strongest in highly compressed members with small load eccentricities. It is fortunate that for such members, typically columns in lower stories of multistory structures for which the axial load capacity may be the most relevant strength parameter, the most liberal criteria b or c may be considered the most appropriate. For more lowly axially loaded members with larger load eccentricities, typically beam-columns and unbraced members for which criterion a is most relevant, effects on capacity of creep are smaller and often insignificant. Partly due to uncertainty of adopted definitions, a wide variety of limits have been proposed in the literature and codes [HEL 05a].

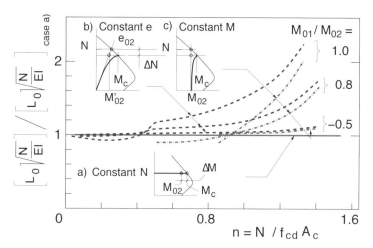

Figure 2.15. *Approximate slenderness limits at 5% capacity reduction according to three criteria (based on the moment multiplier with $C_m = 0.6 + 0.4r_m$ and a N–M diagram for a typical, rectangular section ($\omega = 0.6, h'/h = 0.8, \varepsilon_{yd} = 0.0024$)*

In conjunction with the work leading to EC2, criteria a and b were compared and discussed [WES 04]. The chosen EC2 criterion (10% increase in moments due to second-order effects), can be stated as:

$$M_{Rd} = M_{0Rd} \cdot 1.1$$

which corresponds to about 9% moment capacity reduction ($M_{0Rd}/M_{Rd} = 0.909$).

2.6.2. EC2 – local lower slenderness limits

2.6.2.1. *Code provisions*

According to EC2:2004 (5.8.3.1), second-order load effects in isolated members can be ignored if:

$$\lambda < \lambda_{lim} \tag{2.59}$$

where

$$\lambda_{lim} = \frac{20\,A\,B\,C}{\sqrt{n}} \tag{2.60}$$

$$A = \frac{1}{1 + 0.2\phi_{ef}} \quad , \quad B = \sqrt{1 + 2\omega} \quad \text{and} \quad C = 1.7 - r_m$$

Here,

$\lambda = L_0/i$ is the mechanical slenderness;

ϕ_{ef} is the effective creep factor;

$n = N_{Ed}/f_{cd}A_c$ is the relative normal (axial) force;

$\omega = f_{yd}A_s/f_{cd}A_c$ is the total mechanical reinforcement ratio;

A_s is the area of the total reinforcing steel, and A_c that of the concrete section;

$i = \sqrt{I_c/A_c}$ is the radius of gyration of the area A_c;

and

$$r_m = \frac{M_{01}}{M_{02}} \tag{2.61}$$

is the moment ratio between the smallest, M_{01}, and largest, M_{02}, first-order end moments (including imperfections). The ratio is to be taken positive when the end moments give tension on the same side of the member (single curvature), and negative otherwise (double curvature).

The moment ratio is to be taken as $r_m = 1$ for:

- unbraced members;

- braced members when first-order end moments are due only to, or predominantly due to, imperfections or transverse loading.

2.6.2.2. Comments

Equation [2.60] can alternatively be rewritten as:

$$\lambda^*_{n,lim} = \frac{14\,(2.43 - 1.43r_m)}{1 + 0.2\phi_{ef}} \qquad \text{where} \qquad \lambda^*_n = \lambda\sqrt{\frac{n}{1 + 2\omega}} \quad [2.62]$$

In this form, with λ^*_n being a slenderness normalized with respect to the effect of axial force and reinforcement, it is more suitable for comparison with the NS-EC2 limit below (section 2.6.4).

The first-order moments M_{01} and M_{02}, shall per definition include imperfection effects. For isolated members, these may be included by adding an imperfection eccentricity e_i (see section 2.8). The code does not give any details on how to include $N_{Ed}\,e_i$ in the moment ratio. Should it be added such as to increase the end moments at both ends? Some have argued for this. However, for the end moment ratio, it is most unfavorable to add the imperfection moment $N_{Ed}e_i$ uniformly along the total column length, such as indicated in Figure 2.16, in which the broken (dashed) line represents the first-order moment distribution without imperfections included. This is the approach recommended in this book.

As discussed in section 2.4.3 and defined by equation [2.39], the first-order end moments above should strictly be the second-order sway-modified end moments, M^*_{01} and M^*_{02}. Lateral sway will normally lead to double curvature bending, and thus tend to increase the moment gradient, beyond that caused by gravity loading, and thereby cause smaller r_m values and larger slenderness limits. Neglecting second-order sway effects in the end moment ratio will, therefore, normally be conservative.

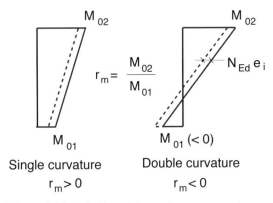

Figure 2.16. *Definition of first-order moment ratio r_m*

2.6.3. NS-EC2 – Local lower slenderness limits

2.6.3.1. *Code provisions*

In NS-EC2:2004 (Norwegian National Annex NA.5.8.3.1), isolated slenderness limits are given in terms of a *normalized slenderness* λ_n defined by

$$\lambda_n = \lambda \sqrt{\frac{n}{1 + 2\,k_a\,\omega}} \quad \text{with} \quad k_a = \left(\frac{i_s}{i}\right)^2 \qquad [2.63]$$

where

λ, n, ω, A_s, A_c are defined previously;

$i_s = \sqrt{I_s/A_s}$ and $i = \sqrt{I_c/A_c}$ are the radii of gyration of the areas A_s of A_c;

I_s and I_c are the second moments of area of A_s and A_c about the centre of A_c.

k_a may simplified (and conservatively) be taken equal to 1.0.

Second-order load effects can be ignored if:

$$\lambda_n \leq \lambda_{n,lim} \qquad [2.64]$$

where

$$\lambda_{n,lim} = 13\,(2 - r_m)A_\phi \qquad\qquad [2.65a]$$

$$r_m = \frac{M_{01}}{M_{02}} \quad , \quad A_\phi = \frac{1.25}{1 + 0.2\phi_{ef}} \leq 1 \qquad\qquad [2.65b]$$

Here,

$r_m = 1$ for unbraced members (free to sway laterally);

$r_m = 1$ for braced members with transverse loading between ends;

$r_m = 1$ if $M_{02} < N_{Ed}h/20$. This restriction applies when end moments are small, in which case uncertainties in moment assessments may significantly affect the end moment ratio.

As pointed out in the previous section, the first-order end moments above should strictly be the second-order sway-modified end moments, M_{01}^* and M_{02}^* (see also section 2.4.3).

2.6.3.2. *Comments*

The limits were derived based partly on elastic analyses and on numerical comparisons with nonlinear material and geometric analyses of reinforced concrete compression members, for a range of different parameters, and different boundary conditions. The presence of a moment gradient delays the development of maximum moment between ends. This is the reason why the limit increases with increasing first-order double curvature bending. In this respect, the limit accounts for the severe case of a member with one end pinned and the other with a strong rotational restraint [HEL 02b, HEL 05a].

Criterion a (10% reduction in moment-capacity) was used, but with an allowance for normal creep at high load levels against criteria b, c. The effect of creep under sustained loading corresponding to about $\phi_{ef} = 1.25$, giving $A_\phi = 1.0$, is considered included in the limit. This covers most practical cases. For more unfavorable cases, A_ϕ becomes less than 1.0, thereby reducing the slenderness limit.

The factor k_a reflects section geometry and reinforcement arrangement. In the detailed derivation [HEL 02b, HEL 05a] of this "section shape/steel arrangement" factor, it was expressed by:

$$k_a = \left(\frac{i_s}{i}\right)^2 \cdot \frac{2.15}{1000\,\varepsilon_{yd}} \qquad [2.66]$$

which also reflects the influence of steel quality. For a B500 steel quality, with design yield stress $f_{yd} = 500/1.15 = 435$ MPa, and yield strain $\varepsilon_{yd} = \sigma_{yd}/E_s = 435/200 = 0.00218$, the last term above becomes close to unity.

Typical values of $k_a = (i_s/i)^2$ are given in Figure 2.17. It can be seen that k_a values of 3–4 are appropriate for many practical cases, and that $k_a = 1$ is always a safe choice, but normally too conservative.

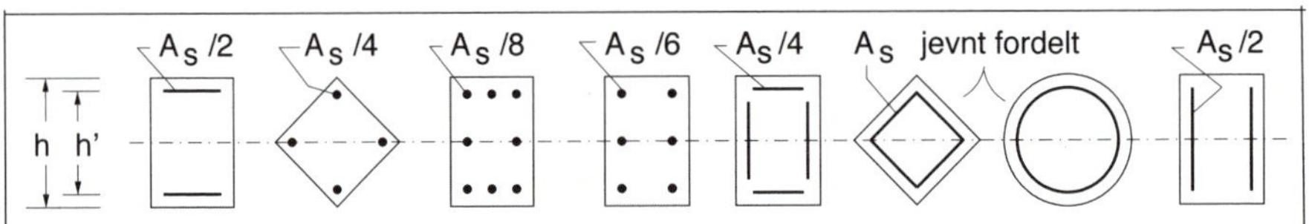

	$A_s/2$	$A_s/4$	$A_s/8$	$A_s/6$	$A_s/4$	A_s jevnt fordelt	$A_s/2$
i/h	0.289	0.204	0.289	0.289	0.204	0.250	0.289
i_s/h'	0.500	0.354	0.433	0.408	0.289	0.354	0.289
i_s/i	1.732 h'/h		1.500 h'/h	1.414 h'/h			1.000 h'/h
$2k_a^{1)}$	4.0		3.0	4.0 x 2/3 = 2.66			1.33
$2k_a^{2)}$	3.0		2.25	3.0 x 2/3 = 2.00			1.00

1) h'/h = 0.82   2) h'/h = 0.71

Figure 2.17. *Radii of gyration and k_a values for typical, symmetrically reinforced, symmetrical cross-section*

2.6.3.3. *Selected comparison with nonlinear analysis results*

Nonlinear reinforced concrete analysis results are shown in Figure 2.18 (from Hellesland [HEL 02b]), for a cantilever column with a moderate first-order moment gradient (see insert).

The EN-EC2 limit for an unbraced column, $\lambda_{n.lim} = 13$, included in the figure, is seen to be in reasonable agreement with the nonlinear results for both low and high reinforcement, and low and high axial load levels. The results were obtained with ideal elasto-plastic corner reinforcement at

$h'/h = 0.8$, and the standard parabola-rectangle stress-strain diagram for concrete with two ultimate strain values (0.003 and 0.0035). The ultimate strain has minor influence on results, as seen.

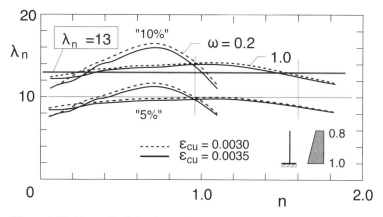

Figure 2.18. *Normalized slenderness limits versus axial load level at 5% and 10% moment capacity reduction, and for two different reinforcement levels*

It is also of interest to consider some results that include creep effects during sustained loading of compression members with high axial forces and small relative load eccentricities (e_{02}/h). It is for such members that creep normally has the most pronounced effect. Some relevant results, for unrestrained and restrained columns, were obtained by Mari and Hellesland [MAR 03, MAR 05b] from numerical nonlinear finite element analyses (FEA) that considered both material and geometrical nonlinearities, cracking and also tension stiffening.

The columns were first subjected to a period (50 years) of sustained loading and then loaded to failure in a short time load application. The load was applied according to each one of the applications defined in conjunction with the three different criteria (a, b, c) discussed previously (section 2.6.1.2). The sustained load fraction ψ was taken, depending on criterion chosen, as a fraction of either the axial load or moment capacities.

Selected 10% capacity reduction results are compared in Figure 2.19 to predictions by the approximate NS-EC limit given by equation [2.65a] for $A_\phi = 1$. The FEA results were obtained for small and very small nominal eccentricities (mostly $e_{02}/h = 0.1$), low and high reinforcement ratios, no concrete strength increase following loading, a sustained load fraction of $\psi = 0.6$, and a linear creep factor of $\phi_{50yrs} = 2$. The latter corresponds to an effective creep factor of approximately $\phi_{ef} = 1.2$.

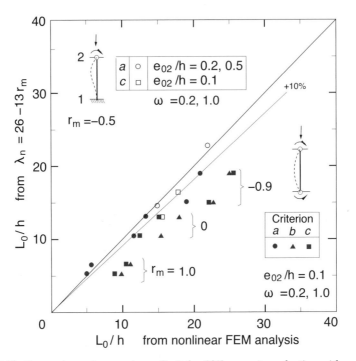

Figure 2.19. *Comparison of approximate limit for 10% capacity reduction with results from nonlinear finite element analyses [MAR 03, MAR 05b] for sustained axial load level = 0.6 and $\phi_\infty = 2$. Criteria: 10% reduction: (a) in moment capacity for constant axial load, (b) in axial load capacity for constant eccentricity and (c) in axial load capacity for constant applied (first-order) moment*

For pinned (unrestrained) columns ($L_0 = L$), the approximate limit is seen to be generally conservative at all considered end moment ratios ($r_m = 1.0, 0, -0.9$). For columns pinned at one end and fully clamped at the other (elastic $L_0 = 0.7L$, $r_m = -0.5$), the approximate limit is just acceptable for results based on criterion a, but there is still ample margin (about 10%) to results based on criterion c, which is the more relevant one at high axial load levels. It may also be noted that a "pinned-clamped" restraint condition is very unfavorable compared to cases with more similar end restraints (as discussed in conjunction with Figure 2.20). A truly "pinned-clamped" restraint condition is also difficult to attain in practice, and is therefore somewhat unrealistic.

These and other results [HEL 02a], document reasonably well that the limit equation [2.65a] allows for normal creep at sustained loads as high as 50%–60% of sectional capacities, and in some cases for significantly higher creep effects.

2.6.4. Comparison of the EC2 and NS-EC2 limits

The introduction of an axial force and reinforcement dependent slenderness, originally labeled "load dependent slenderness" and denoted λ_N [NOR 89], represented a considerable change from previously used slenderness parameters. The label "normalized slenderness λ_n" was introduced in 2002 [HEL 02a]. A significant advantage of this slenderness parameter is that it allows one single number to characterize the combined effect of all main parameters affecting a compression member's "slender" or "non-slender (short)" behavior. It facilitates simplified communication between engineers and other actors in the field. Use of λ, on the other hand, requires further identification of load level and reinforcement.

Although EC2 did not choose to formulate the isolated limits in terms of a normalized slenderness, it can be seen by comparing the EC2 limit expressed by λ_n^* in equation [2.62] with the NS-EC2 limit in equation [2.63], that they contain the same ingredients.

One main difference between the limits is the choice of moment gradient effects. Elastic gradient effects are significantly influenced by the boundary conditions. This is illustrated in Figure 2.20 for columns with two different boundary conditions [HEL 02a]. The full lines are obtained from elastic analyses. Curve (a) is for a member pinned at both ends, whereas curve (b) is for a member pinned at one end (where maximum moment is applied) and with a very stiff ("clamped") rotational restraint at the other end. The latter case is clearly the most unfavorable of the two cases.

The gradient effect of the two code limits are also included in the figure, by the straight lines labeled "1)" for EC2 and "3)" for NS-EC2. The gradient effect of the EC2 limit is close to, or above, the elastic pinned–pinned case a, whereas the NS-EC2 limit is closer, but less conservative, than the "pinned-clamped" case b.

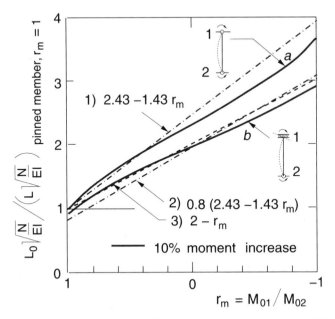

Figure 2.20. *Elastic slenderness limits versus end moment ratio at 10% increase beyond maximum first-order moment, and comparison with some approximations*

Recognizing that real columns may have different restraints at the two ends, the EC2 gradient effect based on the pinned-pinned case seems to be rather too strong. The more prudent NS-EC2 gradient effect seems more appropriate. However, if the comparison is made for a creep factor of $\phi_{ef} = 1.25$, the NS-EC2 limit remains the same, whereas the EC2 limit is reduced to 0.8 times of the short–time value. For this case, the limits are reasonably similar, in particular for r_m values between about 0.8 and -1, as seen by comparing the lines labeled "2)" and "3)". It is advisable, based on these results, to always include creep effects with the EC2 limit.

2.6.5. *Local upper slenderness limit*

Many codes give, in one form or the other, maximum slenderness values that should not be exceeded (summarized in [HEL 90b]). For instance, the

CEB-FIP Model Code 78 [COM 78] gave $\lambda = 140$, or $\lambda = 200$ in conjunction with accurate analysis methods. ACI 318-11 [AME 11] restricts λ in approximate methods to $\lambda = 100$ based on the argument that this value represents an upper value for which experimental results are available.

EC2:2004 does not give any such limits. Although it is the designer's responsibility to ensure sufficient safety for any column slenderness, it is useful to get some indication by a simple means when a column is at the end of its usefulness in terms of slenderness. In this section, some aspects around this topic are considered.

The maximum possible member slenderness for a given axial load $N (= N_{cr})$ is obtained as the one causing buckling under a concentric axial load alone. Such a column has zero first-order moment capacity ($M_{0Rd} = 0$). Its maximum slenderness can be determined from the classical buckling load expression, equation [2.1], in which EI should be taken as the tangent stiffness EI_0 at the origin of the moment-curvature relationship of the reinforced concrete section considered. Solving for the effective length gives:

$$L_{0,max} = \sqrt{\pi^2 EI_0 / N} \qquad \text{and} \qquad \lambda_{max} = L_{0,max}/i \qquad [2.67]$$

A column with such a slenderness is not very useful in practical contexts, as it is not capable of carrying any external end moments or transverse loading.

Figure 2.21 shows computed λ results versus relative axial force n (full lines) for columns with a first-order moment capacity M_{0Rd} that is just sufficient to carry moments due to a specified imperfection, $M_{0Rd} = Ne_i$, and a certain amount of creep. Creep is taken into account in a simplified manner by increasing concrete strains. The results are computed with the total moment given by $M = M_{0Rd} + M_2$ and a moment-curvature relationship obtained with conventional material stress–strain diagram: (1) for concrete, the parabola with the strain $\varepsilon_{c2} = 0.002(1 + \phi_{ef})$ at the peak stress; (2) for reinforcing steel, the bilinear (elasto plastic) diagram with $\varepsilon_{yd} = 0.002$. Two reinforcement ratios are considered: $\omega = 0.2$ is a reasonably representative minimum reinforcement and $\omega = 1.0$ a practical maximum reinforcement.

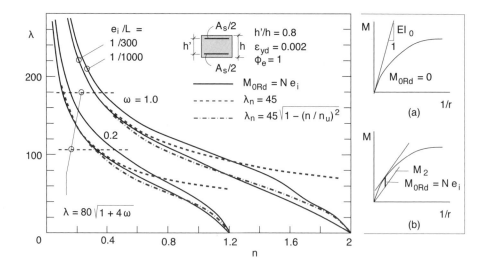

Figure 2.21. *Upper slenderness limit result (from [HEL 90b])*

Based on such results, Hellesland proposed (in 1987) two upper limit expressions [HEL 90b] that also are shown in the figure (broken lines, and dot-dash lines). The two upper limit proposals can be expressed in terms of the present normalized slenderness (equation [2.63]) by

$$\lambda_{n,upper} = 45 \leq 80\sqrt{n} \qquad [2.68]$$

and

$$\lambda_{n,upper} = 45\sqrt{1 - (n/n_u)^2} \leq 80\sqrt{n} \qquad [2.69]$$

where $n_u = 1 + \omega$. Equation [2.68] was adopted by the previous Norwegian Standard NS 3473:1989 and later editions, not as a requirement, but as a recommendation of a slenderness normally not to be exceeded.

The second limit, equation [2.69], can be seen to be a particularly good approximation to the computed curves with $e_i = L_0/300$, which is an imperfection that is only slightly greater than the minimum local imperfection of $e_i = L_0/400$ specified by EC2.

Normally, columns with normalized slendernesses near the values given above, that only allows for axial loading and a margin for imperfection and some creep, will not be suitable as structural load carrying elements. In practical cases, with first-order moments from external loads and imperfections, and creep of normal size, the design process will most often lead to considerably lower slenderness values. If not, there are reasons to be concerned.

In addition to serve as a "warning sign", and thus guard against possible design errors, upper limits are often quite useful in the preliminary design phase, when the first estimate of sectional dimensions and reinforcement assumptions are made.

2.6.6. *Global lower slenderness limit*

Limits indicating when global second-order effects can be neglected have not been so common as local limits. EC2:2004 gives a lower slenderness for global second-order effects at the base of structures such as multilevel buildings.

It can be assumed that second-order effects are less than 10% of the first-order effects when:

$$F_{V,Ed} \leq k_1 \frac{n_s}{n_s + 1,6} \frac{\sum E_{cd} I_c}{L^2} \qquad [2.70]$$

where $k_1 = 0.31$, or $k_1 = 0.62$ if the bracing members are uncracked in the ULS, and if other values are not given in a National Annex of a country.

Further, $F_{V,Ed}$ is the total vertical load ($= \sum N_{Ed}$) on braced and bracing members, n_s is the number of stories, L is the total building height above the level of moment restraint, E_{cd} is the concrete modulus ($= E_{cm}/\gamma_{cE}$), I_c is the second moment of area of the concrete section of bracing members.

The given expression is considered valid provided the structure is (1) reasonably symmetrical, so that torsional instability is not governing, (2) global shear deformations are negligible, (3) bracing elements are fully fixed at the base, (4) stiffness of bracing elements are reasonably constant along the building height and (5) the total vertical load increases by about the same amount per story.

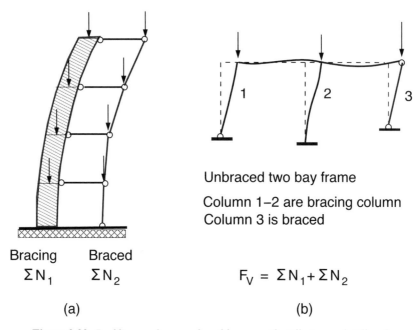

Figure 2.22. *Buckling mode examples of frames with stiffening and stiffened structural elements*

The use of the formulation "global shear deformations are negligible" in the code, is somewhat ambiguous. It seems to be used to exclude cases in which the lateral stiffness is due to flexural sidesway bending on each level, such as illustrated for the unbraced frame in Figure 2.22(b), or such as will be present in cases with shear walls, with or without significant openings.

The limit is clearly developed for building structures such as the one in Figure 2.22(a), from a global bending magnifier formulation, requiring the magnifier to be less than 1.1. Thus, from:

$$M \approx \frac{M_0}{1 - F_{V,Ed}/F_{VB}} \leq 1.1 M_0$$

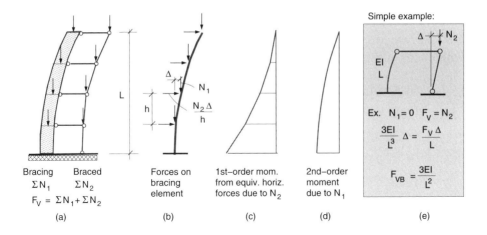

Figure 2.23. *Example of equivalent horizontal forces on vertically loaded frames*

where F_{VB} is the global buckling load, the limiting load can be derived as:

$$F_{V,Ed} \leq (0.1/1.1)F_{VB} \approx 0.1F_{VB} \qquad [2.71]$$

It is reasonably simple to establish approximate buckling loads for such structures, for instance by replacing the effect of vertical loads on deflections by equivalent horizontal loads, such as illustrated in Figure 2.23. The figure also includes a simple frame example, with one bracing and one braced column. From simple equilibrium of the resisting load $(3EI/L^3)$ and the equivalent horizontal overturning load $(F_V \Delta/L)$, the critical load $F_{VB} = 3EI/L^2$ is obtained. Iterative approaches may also be used (see Annex H, for instance).

In EC2, Annex H, the buckling load for multilevel case is given as:

$$F_{VB} = 7.8 \frac{n_s}{n_s + 1,6} \frac{\sum EI}{L^2} \qquad [2.72]$$

This is partly based on curve fitting [WES 04]. For $n_s = 1$, it gives the same load found in the example above.

By assuming EI values of $0.4\,E_{cd}I_c$ and $0.8\,E_{cd}I_c$, for the uncracked and cracked case, respectively, equation [2.70] is obtained from $F_{V,Ed} \leq 0.1F_{VB}$ with the F_{VB} expression above.

For cases with what is denoted significant "global shear deformations" in EC2, Annex H, the annex suggests calculating the buckling load F_{VB} from:

$$F_{VB} = \frac{F_{V,BB}}{1 + F_{V,BB}/F_{V,BS}}$$

[2.73]

This "mixed" formulation can preferably be rewritten as:

$$\frac{1}{F_{VB}} = \frac{1}{F_{V,BB}} + \frac{1}{F_{V,BS}}$$

[2.74]

This would seem to be clearer (and well known as an approximation formulation), and it gives the same result.

$F_{V,BB}$ is the "global buckling load for pure bending" (as defined above, equation [2.72]), and $F_{V,BS}$ is the "buckling load for pure shear". The latter is further defined as $F_{V,BS} = \sum S$, where $\sum S$ is "the total shear stiffness (force per shear angle) of bracing units".

These definitions ("buckling load for pure shear", "total shear stiffness" and earlier, "shear deformations") are again rather ambiguous, as they do not seem to conform to the common definitions related to shear strains. EC2, Annex H, implies that $F_{V,BS}$ is the lateral stiffness of a story, and may involve both flexural and shear strain deformations. For additional details, see [WES 04]. The two effects are considered below.

For shear alone, Hooke's law defines shear strain as $\gamma = \tau/G$, where τ is the shear stress and G the shear modulus given in terms of the elastic modulus and Poisson's ratio ν by $G = E/2(1+\nu) \approx 0.4E$. Then, for a shear wall panel with area A in the shear direction, multiplication with A in the numerator and denominator gives an average shear strain:

$$\bar{\gamma} = \bar{\tau}A/GA = H/S_{shear}$$

where $\bar{\tau}$ is an average shear stress and H is the horizontal force required to give the shear strain $\bar{\gamma}$ of a panel with sectional shear stiffness $S_{shear} = GA$.

Now, only flexural sidesway of a story (single-level frame) is considered. A lateral load H, applied, for instance, to the frame in Figure 2.22(b), will give a relative horizontal displacement Δ_0 (first order) and the inclinations, defined by the chords through the top and bottom of the columns, given by $\gamma_i = \Delta_0/h_i$ for the various columns. For the common case with equal column lengths $h_i = h$, the story inclination will be the same in all axes, that is $\gamma_i = \gamma$. The lateral flexural stiffness of a story per unit inclination, S_{flex}, can then be defined by:

$$S_{flex} = H/\gamma = H/(\Delta_0/h)$$

Total lateral stiffness (per unit inclination) from flexure and shear then becomes:

$$\sum S = S_{shear} + S_{flex}$$

It seems that this is what is implied by the term "the total shear stiffness (force per shear angle) of bracing units" in EC2. The term "lateral stiffness" would have been better.

2.7. Effect of creep deformations

2.7.1. *General*

For a member in flexure, subject to sustained loads, or so-called long term or quasi-permanent loads, displacements along the member will increase with time due to time-dependent creep deformations. For statically determinate members with axial forces, such as cantilever or simply supported columns, increased deformations will result in increased moments, and as a consequence, reduced ability of the member to carry external loads. In statically indeterminate frames, the effects of the time-dependent creep deformations are more complicated, but they will lead to a redistribution of moments. This may cause moment increases in some portions and decreases in other portions. For instance, a moment relief may take place in columns restrained by stiff beams. The net effect may be an increase in the frame's load carrying capacity (shown, for instance, in [MAN 67]). In practical design, it is not such cases, but rather cases with reduction in load carrying capacity due to creep that are of most concern.

Simultaneously with the development of creep deformations with time, the concrete strength will generally increase due to continued hardening (hydration). In some instances, this strength increase is sufficient to cancel detrimental creep effects. Hardening effects are dependent on the type of cement used, age at loading (t_0), environmental conditions and sustained stress levels. Normally, not least in standard design, continued hardening effects are conservatively neglected due to the uncertainties of their magnitude.

Creep effects depend on the long-term load history during the structure's lifetime. The detailed history may be quite complicated, and it is therefore necessary to simplify. Codes generally define creep to be due to serviceability loads (without load factors). A typical design calculation history may look like that illustrated in Figure 2.24(a) in terms of axial force versus mid-height

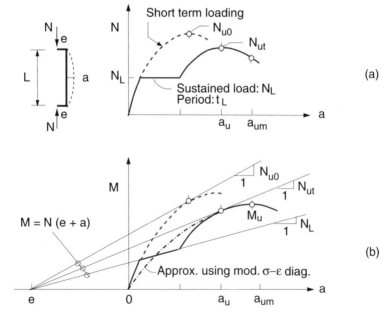

Figure 2.24. *Typical design load history in terms of (a) load–deflection diagram and (b) moment–deflection diagram*

deflection for a simply supported column with a constant initial eccentricity (e). The same results are shown in Figure 2.24(b) in terms of moment versus deflection at the critical (mid-height) section.

In the figure, N_{u0} is the ultimate load capacity (limit load) as obtained under short-time loading, and N_{ut} is that obtained under short-time loading following the sustained (long-term) load period t_L.

If the ultimate loads are reached (at a deflection a_u) before the maximum moment capacity (M_u) of the critical section is reached (at a_{um}), such as is seen to be the case in the illustration of Figure 2.24(b), the failure mode is said to be due to instability. In other cases, section capacity may be limited by material failure (at defined failure strains) prior to column instability.

2.7.2. *Effects on load and deformation capacity*

2.7.2.1. *Load capacity*

The ratio N_{ut}/N_{u0} gives the effect of creep on the load carrying capacity. Samples of such results, taken from Hellesland [HEL 70a, HEL 70b], are shown in Figure 2.25. They were obtained for simply supported, constant eccentricity columns with a rectangular section and symmetrically placed corner reinforcement with ideal elasto plastic stress–strain properties. The ascending and descending branches of the instantaneous, or short-time, concrete stress–strain diagram are illustrated in the insert in Figure 2.25. The curve was approximated by a nonlinear exponential relationship with a strain of ε_{c1} at the peak (strength).

Figure 2.25 shows the selected results as function of the sustained load level N_L/N_{u0}. They were obtained using a rather rigorous theory based on a strongly nonlinear creep-stress relationship and a "modified rate of creep" method, that allows for creep recovery and provide greater creep strains than the regular "rate of creep" method (for review of different creep theories, see, for instance, [BAZ 03]). Solutions for strains and stresses at a number of "fibers" across the critical section, and deflections for increasing time, were obtained using an incremental, stepwise propagation procedure. The concrete material law included both detrimental effects of high sustained stresses and beneficial effects of concrete strength increase due to continued hardening (hydration) following loading [HEL 70a, HEL 72]. Shrinkage, preloading strains due to shrinkage and concrete tensile strength were also included, but had minor effects.

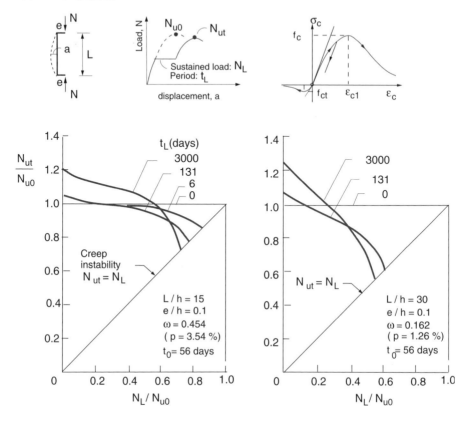

Figure 2.25. *Effect of sustained loading on load capacity of rectangular, corner reinforced, constant eccentricity columns; $h'/h = 0.65$, $\varepsilon_y = 0.193\%$, $\varepsilon_{c1} = 0.225\%$, $\phi_{\infty,56d} = 2.3$ in the linear range and ≈ 3.1 for $\sigma_c/f_c = 0.5$, $/f_{c,\infty} = 1.30 f_{c,56d}$ (from [HEL 70a, HEL 70b])*

The columns in the figure were loaded at a concrete age of $t_0 = 56$ days. Between this age and infinity, taken to be about 3,000 days, continued hydration increased the concrete strength by a reasonable modest 30%. The nonlinear creep-stress law adopted, yielded for this case a creep factor of about $\phi_{\infty,56} = 2.3$ at lower stress levels (linear creep-stress range), and about $\phi_{\infty,56} = 3.1$ at a concrete stress level of about $\sigma_c = 0.5 f_c$.

At lower sustained load levels, it is seen that the concrete strength increases more than cancels the detrimental effects of creep, thus giving N_{ut}/N_{u0} values greater than 1.0. At higher sustained load levels, the opposite is the case. In particular for the most slender column ($L/h = 20$), the decrease in load capacity is very significant. The two columns considered in the figure both have a small eccentricity of $e/h = 0.1$. Detrimental creep effects is particularly strong for end eccentricities of about this magnitude. By neglecting concrete strength increase due to continued hydration, the curves in the figures will be lowered somewhat.

If the sustained load level N_L/N_{u0} is sufficiently high, the column may fail during sustained loading. This phenomenon is referred to as *creep instability* (or creep buckling). This situation, in which the ultimate load is equal to the long-term load ($N_{ut} = N_L$) is obtained at the intersection between the capacity curves and the 45 degree lines in Figure 2.25. Creep instability is a situation we want to avoid by including creep effects in the design.

Figure 2.26 shows sustained creep load levels versus mid-height deflection for various sustained load periods. The load that causes creep instability at $t_L = \infty$ (here approximated by $t_L = 3,000$ days) is denoted as the *sustained load capacity*. For sustained loads below this level, creep effects level off with time and the deflection terminate at finite values. In such cases, there is a reserve load capacity at the end of the sustained load period. For sustained loads above the sustained load capacity, however, the column will fail due to creep instability at finite times.

Detrimental effects on load capacities decrease with increasing end eccentricities and increasing reinforcement, and increase with increasing slenderness. Results demonstrating this, in the absence of hydration effects, have been presented by several investigators, including K. Aas-Jakobsen [AAS 73] in an earlier study using finite element analyses. Some of his results are also presented in CEB-FIP Buckling Manual [COM 77]. Selected results based on the "rate of creep method" are shown in Figure 2.27. This method gives somewhat smaller creep effects than the "method of superposition", the "strain hardening method" and the "reduced modulus" method (using a modified concrete stress–strain diagram), which also have been presented by Aas-Jakobsen.

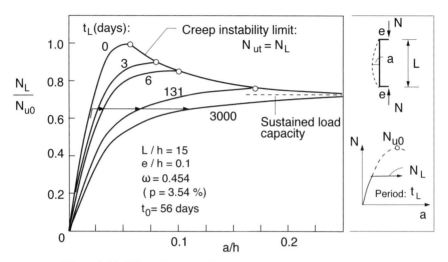

Figure 2.26. *Effect of sustained loading on load capacity; data as in Figure 2.25 (from [HEL 70a, HEL 70b])*

The geometric slenderness that corresponds to the slenderness limit $\lambda_{n,\lim} = 13 A_\phi$ (equation [2.65a]) is also plotted in the figure. It becomes $L_0/h = 2L/h = 9.2$ for $\phi_{ef} = 2.2/1.5 = 1.47$ and $A_\phi = 1.25/(1 + 0.2\phi_{ef}) = 0.97$. It can be seen that creep effects are small, and negligible for columns close to the lower slenderness limit in this case, as in many other practical cases.

2.7.2.2. Deformation capacity

In the previous section, the effects of creep on load capacities have been considered. In some cases, the effect of creep on columns' deformation capacities may be as important. For instance, in long bridges, in which the superstructure (beam, girder) is generally considerably stiffer than the columns. It is quite often designed as simply supported on column supports, and its rotations and displacements at a column support are normally little influenced by the stiffnesses of integrally connected columns. Then, as far as the columns are concerned, the main actions that should be considered in addition to the axial forces are the deformations (rotation, displacement) imposed by the stiff superstructure to the column tops. The associated deformation induced column moments and shears are relatively proportional to the stiffness of the selected columns, and are of secondary interest in such cases. For design for imposed deformations, see also section 2.9.1.

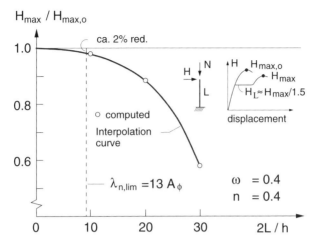

Figure 2.27. *Effect of creep on the load capacity of transversely loaded columns; $h'/h = 0.8$, $t_L = 700$ days, $\varepsilon_y = 0.21\%$, $\phi_{100d} = 1.25$, $\phi_{700d} = 2.2$, $\phi_\infty = 2.5$ (from [AAS 73])*

The displacement and rotation capacity at an end of a column are affected by the length of time taken to apply the imposed deformations, and of the column slenderness (stability effects). Temperature effects will normally contain a short-term (daily) element and a long-term (seasonal) element. Long-term effects may also be due to creep in the superstructure caused by prestressing forces. Deformations imposed over a long-term will allow creep in the columns to reduce the associated moments and shears, and normally increase a column's displacement and rotation capacity. An opposite effect is that of creep in the columns that will cause increased deflection and second-order (stability) effects between ends. This may limit the deformation capacity by premature column instability. Studies of columns subjected to imposed deformations, such as [HEL 81] and [HEL 85], may be consulted to obtain a better understanding of the topic.

2.7.3. *Approximate calculation of creep effects*

In addition to causing increases in displacements and moments, creep also results in stress redistribution from the concrete in the compression zone to the compression reinforcement, and may cause premature yield of the compression steel. Exact calculations, even for the simple load histories, are quite complicated, and require access to suitable computer programmes.

Several such programmes, generally finite element based, are available today, but not used in normal design calculations. Approximate procedures are presented below.

2.7.3.1. *Effective creep*

Consider first a concentrically loaded pure concrete specimen in the low stress range, with $\sigma = E\varepsilon$ governing the short-time (instantaneous) response, and with a unit creep (creep per unit stress) given by ϕ. For a stress history consisting of a short-time application of σ_L, a subsequent sustained period with this stress and finally a short-time stress increase to σ_D, the total strain $\varepsilon_{D,tot}$ corresponding to σ_D, can in simplified manner be written as:

$$\varepsilon_{D,tot} = \frac{\sigma_L}{E} + \frac{\sigma_L}{E}\phi + \frac{\sigma_D - \sigma_L}{E} = \varepsilon_D(1 + \phi_{ef}) \qquad [2.75a]$$

where

$$\varepsilon_D = \frac{\sigma_D}{E} \qquad \text{and} \qquad \phi_{ef} = \phi\frac{\sigma_L}{\sigma_D} \qquad [2.75b]$$

are the total short-term strain for σ_D and "the effective creep coefficient", respectively. With such a creep factor, we avoid having to consider the load history explicitly. In EC2, the ϕ factor above is in design contexts taken as ϕ_{∞,t_0}, which is the final unit creep coefficient at $t = \infty$ for an age of loading t_0.

In a specific case where the sustained load period t_L is less than $t = \infty$, for instance, when comparing with test results, it would seem reasonable to use ϕ_{t_L,t_0} for the actual $t = t_L$.

At higher concrete stresses, creep increases with stress in a nonlinear fashion. In such cases, EC2 suggest that the final creep coefficient above (ϕ, equation [2.75b]) should be replaced by:

$$\phi_{k(\infty,t_0)} = \phi_{(\infty,t_0)}\, e^{1.5(k_\sigma - 0.45)} \qquad \text{where} \qquad k_\sigma = \sigma_c/f_{cm,t_0} \qquad [2.76]$$

in order to reflect nonlinear creep. According to this relationship, which implies a rather modest creep nonlinearity, the transition from linear to nonlinear creep is at a concrete stress of $\sigma_c = 0.45 f_{cm,t_0}$ where f_{cm,t_0} is the mean concrete compressive strength at the time of loading.

Also for the complete structure, creep effects may be accounted for in terms of an effective creep coefficient ϕ_{ef} that accounts in a simplified manner for the long-term, quasi-permanent load action. In the linear creep–stress range, it is defined by:

$$\phi_{ef} = \phi_{(\infty,t_0)} \frac{M_{0Eqp}}{M_{0Ed}} \qquad [2.77]$$

in EC2. Here, the subscript qp refers to long-term loads, termed "quasi-permanent loads" in the EC2. Furthermore, M_{0Eqp} and M_{0Ed} are the first-order bending moments for the long-term loads in SLS and for the design loads in ULS, respectively.

Westerberg [WES 04] has shown that this moment ratio in the ϕ_{ef} definition gives a good approximation of creep effects at the section level of both plain and reinforced concrete in the linear material stress–strain range. In a real structure, moments should strictly be total moments (including second-order effects) rather than first-order moments. This would require iterations, and would be too cumbersome.

A complication with ϕ_{ef} given in terms of moment is that moments vary from section to section along a member. Simplified, the section with the maximum moment may be used according to EC2, or a representative mean value that better represents the total member.

In the CEB-FIP Model Code 1978 [COM 78], the effective creep coefficient was defined by:

$$\phi_{ef} = \phi_{\infty,t_0} \frac{N_{Eqp}}{N_{Ed}} \cdot \frac{M_{Eqp}}{M_{0Ed}} \qquad [2.78]$$

as a function of the sustained axial load ratio (N_{Eqp}/N_{Ed}) in addition to the sustained moment ratio. This formulation will give smaller effective creep coefficients than that in EC2:2004. In the CEB-FIP Model Code 1990 [COM 93], a corresponding formulation is given, but without ϕ_{∞,t_0} included in the ϕ_{ef} definition.

A simpler approach than that given by equation [2.77], but possibly not as well argued theoretically, would have been to take ϕ_{ef} proportional to the ratio between the applied loads rather than resulting moments. Such a definition was adopted, for instance, by Aas-Jakobsen [AAS 73], and in a different form by the ACI 318 code [AME 11].

2.7.3.2. *Approximate method using modified concrete $\sigma_c - \varepsilon_c$ diagram*

The effect of creep can in a simplified manner be obtained by replacing the short-time strain ε_c in the concrete stress–strain diagram by:

$$\varepsilon_c(1 + \phi_{ef}) \tag{2.79}$$

also in nonlinear material and geometric structural analyses. This approach is well known, and it has been considered by several investigators, including Aas-Jakobsen [AAS 73]. It is also discussed in the CEB-FIP Buckling Manual [COM 77]. In lieu of more accurate methods, this approach is also accepted by EC2:2004 (5.8.6) in the context of "the general analysis method", that considers both nonlinear material and geometric effects. Differences in the various applications in the literature are generally only due to differences in the ϕ_{ef} definition.

In Figure 2.24(b), the dash–dot line indicates what is hoped to be achieved with a creep-modified stress–strain diagram. The method is very simple and can easily be incorporated into existing computer programs, including such programs or hand calculations for section design.

2.7.3.3. *Alternative simplified methods*

Also in the two simplified methods in EC2, creep is accounted for by use of the effective creep coefficient, either to reduce the section stiffness or to increase the curvature. The correspondence between the two approaches is briefly considered here, and an effort is made to try to explore the specific formulations.

A consequence of the modified stress–strain diagram, equation [2.79], is that the elastic modulus is reduced. The approach described above may therefore also be denoted as the "reduced modulus approach". The reduced concrete modulus in the approach above becomes:

$$E_c/(1 + \phi_{ef}) \tag{2.80}$$

where E_c is the short-time value. The reduced modulus is suitable for inclusion in moment magnifier-type methods. It is adopted in the so-called "Nominal stiffness" method in EC2:2004 (5.8.7) as defined by the nominal section stiffness in equation [2.108]:

$$EI = \frac{K_{c0}}{1 + \phi_{ef}} E_{cd}I_c + E_s I_s \qquad [2.81\text{a}]$$

where

$$K_{c0} = n\frac{\lambda}{170}\sqrt{\frac{f_{ck}}{20}} \leq 0.2\sqrt{\frac{f_{ck}}{20}} \qquad [2.81\text{b}]$$

K_{c0} accounts for effects of concrete cracking, and is, it is believed, calibrated to the general method results for various concrete grades (strength classes f_{ck}), relative axial forces (n) and slendernesses (λ). In comparison, in ACI 318-2011, K_{c0} is simply taken as $K_{c0} = 0.2$.

EI in equation [2.81a] can, for the present purpose, be written as:

$$EI = \frac{EI_0}{1 + \phi'_{ef}} \qquad [2.82]$$

where $EI_0 = K_{c0}E_{cd}I_c + E_s I_s$, and where ϕ'_{ef} is a modified effective creep coefficient that can be solved for from equation [2.82] and expressed by:

$$\phi'_{ef} = \frac{\phi_{ef}}{1 + \frac{E_s I_s}{K_0 E_c I_c}\phi_{ef}} \qquad [2.83]$$

The short-time stiffness EI_0 and the reduced stiffness are illustrated in the moment-curvature diagram in Figure 2.28. Equilibrium between external and internal forces is obtained at points A and B in the two cases.

This implies that the equivalent approach in terms of curvatures would be to replace the short-time (instantaneous) curvature $1/r_0$ by an increased value:

$$\frac{1}{r} = \frac{1}{r_0}K_\phi \qquad \text{where} \qquad K_\phi = 1 + \phi'_{ef} \qquad [2.84]$$

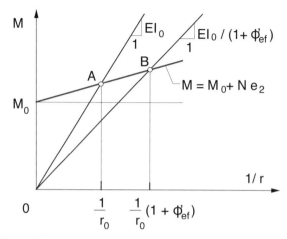

Figure 2.28. *Relationship between creep-modified stiffnesses and corresponding curvatures*

In the "Nominal curvature method" in EC2, the creep effect is included in the nominal curvature (equation [2.115a]) through a multiplication factor defined by:

$$K_\phi = 1 + \beta\phi_{ef} \geq 1 \qquad\qquad [2.85a]$$

where

$$\beta = 0.35 + f_{ck}/200 - \lambda/150 \qquad\qquad [2.85b]$$

It is seen that this β factor accounts for concrete grade (strength class f_{ck}) and slenderness (λ). And probably also cracking, reflected through the constant 0.35.

No explanation is given in EC2, or found elsewhere, on how this β factor is derived. However, the derivation of ϕ'_{ef} above provides a good base for the curvature formulation. In comparison with equation [2.83], it is clear that $\beta\phi_{ef}$ and ϕ'_{ef} are functions of some of the same parameters. Possibly, β is found by calibration to "the general method" results.

In the EI expression, equation [2.81a], only the concrete contribution is reduced for creep effects. An effect of creep is also to cause a stress transfer from the concrete in compression to the compression steel, and thus induce premature yielding, and consequently reduced stiffness. It could therefore be argued that also the reinforcement contribution to the stiffness should be somewhat reduced for creep. The ACI-318 code [AME 11] employs such an approach.

In the literature, several alternative simplified approaches can be found. One approach is to include creep through a creep eccentricity, through a Dischinger formulation of the kind used in the CEB-FIP Model Code from 1978 and 1990 [COM 78, COM 93]. Although in principle rather simple, it is more cumbersome than the two alternative approaches adopted by EC2:2004.

Different simplified methods will reflect somewhat different creep effects. However, as the creep contribution to the total load eccentricity normally is small, it will not matter too much if we use one or the other method.

2.7.3.4. Neglect of creep effects

According to EC2 (5.8.4) effects of creep can be ignored when the three following conditions are satisfied:

$$\phi_{\infty,t_0} \leq 2 \quad , \quad \lambda \leq 75 \quad \text{and} \quad M_{0Ed}/N_{Ed} \geq h \qquad [2.86]$$

where ϕ_{∞,t_0} is the final creep factor for a given age at loading (t_0), M_{0Ed} and N_{Ed} are corresponding first-order moment and axial force, respectively, in the considered (ULS) load combination. These are the same limits given in the CEB-FIP Model Codes.

2.7.3.5. ACI 318 approach

The stiffness formulations above have clear similarities with the corresponding ACI 318 code [AME 11] formulations given by:

$$EI = \frac{0.2\,E_c I_c + E_s I_s}{1 + \beta_d} \quad \text{or} \quad EI = \frac{0.4\,E_{cd} I_c}{1 + \beta_d} \qquad [2.87]$$

where β_d is a factor (less than 1.0) that accounts for creep effects. In the case of braced (non-sway) columns, $\beta_d = \beta_{dns}$, defined as the ratio between the maximum sustained axial load to the maximum axial load in the load combination, both calculated with ULS load factors. For sway action,

$\beta_d = \beta_{ds}$, defined as the the maximum sustained shear (lateral load) to the maximum shear in the load combination, both calculated with ULS load factors.

2.8. Geometric imperfections

2.8.1. *Imperfection inclination*

According to EC2 (5.2), it is necessary, except in SLS calculations, to consider effects of possible unfavorable effects of deviations from the intended   structural   geometry   (inclinations   and/or   preloading out-of-straightness) and of theoretical load positions. Such imperfections, or unintentional eccentricities, must be taken into account for both slender and non-slender (short) structural elements.

For members with axial compression, typically columns, and frames with vertical loads, imperfections can, according to EC2, be represented by an inclination (from the vertical) defined by θ_i (subscript i for imperfection) given by:

$$\theta_i = \theta_0 \cdot \alpha_h \cdot \alpha_m = \frac{1}{200} \cdot \frac{2}{\sqrt{L}} \cdot \sqrt{0.5 \left(1 + \frac{1}{m}\right)} \qquad [2.88]$$

where $\theta_0 = 1/200$ (recommended) is the basic value, $\alpha_h = 2/\sqrt{L}$ is a reduction factor for height (length) limited by $2/3 \geq \alpha_h \leq 1$, $\alpha_m = \sqrt{0.5\,(1 + 1/m)}$ is a reduction factor for number of members, L (in meters) is the height (length) of the structure, or structural element, and m is the number of structural elements contributing to the total effect. The factors α_h and α_m clearly reflect probability aspects.

This imperfection is a "first-order quantity". Load eccentricities due to this imperfection are considered to be a part of the first-order load eccentricities (e_0), or moments (M_0). Additionally, for slender structures and structural elements, there will be second-order effects of the imperfections, in the same manner as for other first-order effects.

2.8.2. *Stiffening structural elements*

Imperfections can be accounted for by including them in the geometry of the analysis model. Generally it will be more convenient, however, to base the analysis on the perfect geometry model (with no imperfections) and to account for imperfections through equivalent external loads. This is particularly so for imperfection effects in structural elements that provide the lateral bracing, in total or in part, of a structure.

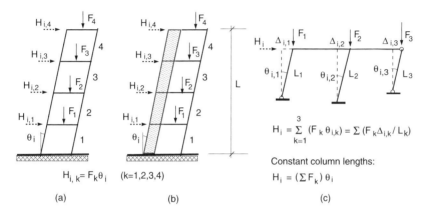

Figure 2.29. *Imperfections in (a) multilevel frame, (b) multilevel frame with a shear wall and (c) multibay frame*

Three examples are shown in Figure 2.29 for the sidesway action in frames. The effect of vertical loads on an inclined structure will cause overturning moments that can be replaced by equivalent horizontal forces. In the multilevel frame, Figures 2.29(a) and (b), the imperfection can be accounted for by applying a horizontal load at the top of each level that is given by

$$H_i = F_k \Delta_i / L_{story} = F_k \theta_i \qquad k = 1, 2, 3, ... \qquad [2.89]$$

where F_k is the total floor load.

Thus, imperfections are conveniently accounted for by adding H_i to the other load cases. In first-order analysis for such load cases, imperfection effects will be included in the first-order results. If instead, a second-order analysis had been carried out, the second-order effects of the imperfections will also be included. It is emphasized that the equivalent horizontal forces are fictitious and that the corresponding shear forces are not real. Therefore, they must be deducted from the total shear forces caused by the real loads.

Example 1: For the multilevel moment frame in Figure 2.29(a), the columns at each floor level provide the lateral stiffness required to maintain sidesway stability. With four stories of 3 m each, $L = 12$ m, and $\alpha_h = 2/\sqrt{15} = 0.52$. In this case, where $m = 8$ columns (vertical members) contribute to the total imperfection effect the imperfection inclination becomes:

$$\theta_i = \theta_0 \, \alpha_h \, \alpha_m = \frac{1}{200} \cdot \frac{2}{\sqrt{12}} \cdot \sqrt{0.5 \left(1 + \frac{1}{8}\right)} = \frac{1}{200} \cdot 0.577 \cdot 0.750 = \frac{2.2}{1000}$$

Example 2: For the similar multilevel frame in Figure 2.29(b), lateral stability is provided mainly by a distinct bracing system consisting of a shear wall, and, to a minor extent, by continuous columns. If the columns had had hinge-like end supports, the shear wall would have to provide the total required lateral stiffness and strength. Even in cases with continuous columns, the shear wall will normally be designed to provide the total lateral stiffness and strength, since the lateral stiffness contribution from the columns most often is negligible compared to that of the shear wall. Columns can in such cases be considered braced, and must additionally be considered for imperfection effects due to out-of-straightness, etc., as discussed below for isolated members, and illustrated in Figure 2.30(b) for the pinned-ends column 2.

For the imperfection of bracing systems, EC2 (5.2(6)) states that m should be taken as the number of vertical members that contribute to the horizontal forces on the bracing system, that is $m = 4$ in this case. With this value, the inclination becomes $\theta_i = 2.3/1,000$, which is only slightly greater than the value for the moment frame above. Uncertainties with respect to the calculation of m will normally have a minor effect on computed imperfection.

Example 3: In the multi-bay frame in Figure 2.29(c), the equivalent horizontal imperfection loads could have been applied at the top of each column. However, for a reasonable inextensible top beam (girder, floor), which is in practice a reasonable assumption, the sum of these forces can be applied at one point at the beam level, as illustrated in the figure by H_i.

The column inclinations will be different if the column lengths are different in the various axes. An assumption to the conservative side would be to compute θ_i in each axis from equation [2.88] with $m = 1$ (as for an isolated column) and to compute the resulting H_i as indicated in the figure. However, considering the approximate nature of the inclination expression (equation [2.88]), it seems reasonable for simplicity to take L equal to the average length (height).

For an average value of $L = 5\,\mathrm{m}$, and with $m = 3$ (columns):

$$\theta_i = \theta_0\, \alpha_h\, \alpha_m = \frac{1}{200} \cdot \frac{2}{\sqrt{5}} \cdot \sqrt{0.5\left(1 + \frac{1}{3}\right)} = \frac{1}{200} \cdot 0.894 \cdot 0.816 = \frac{3.7}{1000}$$

Example 4: The one-bay frame in Figure 2.30(a) is included to illustrate the interaction between stiffening and stiffened columns. Column 1 is clearly a stiffening column that provides lateral bracing to column 2, pinned at both the ends. The equivalent horizontal force due to the imperfection inclination is given by:

$$H_i = \sum P_k \Delta_i / L = \sum P_k\, \theta_i \qquad k = 1, 2 \qquad\qquad [2.90]$$

For the case with large beam stiffness ($\approx \infty$), $M_{i,A} = M_{i,B}$. Then, noting that $N_1 = P_1$, the imperfection eccentricity for column 1 becomes:

$$e_{i,B} = \frac{M_{i,B}}{P_1} = \frac{\Delta_i}{2}\left(1 + \frac{P_2}{P_1}\right) \qquad\qquad [2.91]$$

This $e_{i,B}$ also includes the effect of the inclination of column 2. The corresponding e_i in column 2 from inclination effects are zero, since the column does not contribute to the lateral resistance. First-order imperfection moments, and maximum imperfection eccentricity (at B), are shown in Figure 2.30(b).

The effect of a possible out-of-straightness imperfection between ends need not be considered for the stiffening column 1. For the braced column 2, however, this effect must be considered as illustrated in the figure (by a sinusoidal distribution) and as described for isolated columns below.

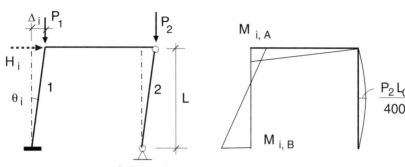

Stiffening
(bracing)
column

Stiffened
(braced)
column

Moments and eccentricity

Column 1 $e_{i, B} = M_{i, B} / P_1$

(a)   (b)

Figure 2.30. *(a) Portal frame with a stiffening and stiffened column, (b)
first-order imperfection moments and eccentricities*

2.8.3. *Stiffened and isolated structural elements*

Application of an equivalent imperfection force is also a suitable approach for isolated compression members (typically columns), whether statically determinate or indeterminate.

Consider, for instance, a cantilever column of length L and vertically loaded by P at the top. The equivalent horizontal imperfection load at the top becomes $H_i = P\Delta_i/L = P\theta_i$, and the moment at the base $M_i = H_i L = P\theta_i L = P\theta_i L_0/2$. The imperfection load eccentricity $e_i = M_i/N$ then becomes:

$$e_i = \theta_i \cdot \frac{L_0}{2} \qquad\qquad [2.92]$$

where L_0 is the effective length. By conservatively taking $\alpha_h = 1$, and $m = 1$, giving $\alpha_h = 1$, the inclination becomes $\theta_i = \theta_0 = 1/200$. Then:

$$e_i = \frac{L_0}{400} \qquad\qquad [2.93]$$

represents a conservative imperfection eccentricity. The same expressions can be derived for other cases.

For braced columns, the imperfection will mainly be due to out-of-straightness (initial, preloading curvature distribution) between column ends, and due to deviations in vertical load positions (end eccentricities) from the intended position. Both of these effects are accounted for by the imperfection eccentricity above.

2.9. Elastic analysis methods

2.9.1. *Principles, equilibrium and compatibility*

When applying fully nonlinear analysis, analysis and proportioning control are carried out in one step. In practical applications, however, use of the fully nonlinear analysis method is generally not convenient. As an alternative, and simpler, second-order elastic analysis may be used. It represents a two-step procedure, in which first- and second-order results are computed in one and the same first step, and proportioning of the members is carried out in the second step. First-order elastic analysis may be used when second-order effects are negligible. First-order and second-order elastic analysis computer programmes are widely used in many contexts. Its application to reinforced concrete structures is just barely mentioned in EC2. It is briefly presented here, also for the purpose of discussing principles of equilibrium and compatibility. The basic steps of such a two-step analysis can be summarized as follows:

1) Assume secant stiffnesses for the reinforced concrete members that reflect cracking and axial forces, evaluate possible long-term creep effects, etc., carry out the second-order elastic analysis based on these secant stiffnesses.

2) Proportion members such that resulting stiffnesses become approximately equal to assumed stiffnesses at a reasonable number of sections along the members. In this manner, both equilibrium and compatibility will be satisfied at those sections.

The main problem with such methods is that the analysis is elastic while the detailed design (proportioning) involves internal sectional moment resistances (capacities, strengths, (M_{Rd})) that are functions of nonlinear constitutive stress–strain relationships, cracking, etc. Nonlinear, internal moment-curvature relationships, such as those shown by the nonlinear curves in Figure 2.31 are dependent on axial force, reinforcement magnitude and placement and cross-sectional dimensions.

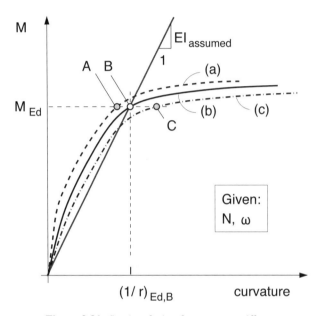

Figure 2.31. *Section design for constant stiffness*

In the first step, it is primarily the bending secant stiffness (EI) that requires attention. For an assumed section with a given axial force and reinforcement, a typical, assumed secant stiffness is illustrated in Figure 2.31 together with the internal moment-curvature relationship. In statically indeterminate structures, also the axial stiffnesses (EA) affect results. Normally, however, such effects are small, and most often negligible.

Now, if the second-order elastic analysis gives an external moment M_{Ed} at a specific section, the corresponding external curvature becomes $(1/r)_{Ed} = M_{Ed}/EI$. A design solution that would satisfy both equilibrium and kinematic compatibility in terms of curvature is then given by the reinforcement that provides an internal moment-curvature resistance curve that passes through the equilibrium point B in Figure 2.31. This curve is labeled (b) in the figure.

Curve (c), obtained with less reinforcement than curve (b), does not satisfy equilibrium at the external curvature $(1/r)_{Ed}$, but at a larger curvature at point

C, corresponding to a smaller stiffness (giving larger second-order effects) than assumed in the analysis. This solution is not acceptable (seen in isolation for the considered section).

Curve (a), on the other hand, based on a higher amount of reinforcement than curve (b), provides equilibrium at a smaller curvature, at A. Curve (a) represents a satisfactory solution, but with more reinforcement than necessary. As a result, the section has a larger stiffness than the section used in the analysis. If this is also the case at other sections, member displacements and second-order effects will be smaller than those computed in the linear second-order analysis. A re-analysis with a revised stiffness assumption may be considered.

2.9.2. Equilibrium and compatibility at multiple sections

A reasonably strict design would require that a number of sections along the member are proportioned such that both equilibrium (between external and internal forces) and kinematic compatibility (between external curvatures (M_{Ed}/EI) and internal curvatures $(1/r$ for the internal moment equal to $M_{Ed})$ are satisfied.

To illustrate this further, consider the cantilever column shown by the inset in Figure 2.32. Also shown is the total moment diagram (first- plus second-order effects) obtained by an elastic second-order analysis based on an assumed section stiffness EI, constant along the column. Instead of a constant EI, we could have assumed a stepwise constant stiffness. This would not change the basic principles involved.

The column is divided into three elements. For an assumed section size, the reinforcement (ω) is proportioned for the maximum moment within each element. In order to satisfy both equilibrium and compatibility, the necessary reinforcement amounts at each of these sections are those (A_{s1}, A_{s2}, A_{s3}) providing the internal $M - 1/r$ curves (dashed lines) passing through the illustrated external equilibrium points.

By selecting stepwise constant reinforcement in the manner described above, the reinforcement will be greater than required at all but the three considered sections. As a result, the column becomes stiffer than reflected by the assumed stiffness, the external moments become smaller and the design becomes conservative (to the safe side).

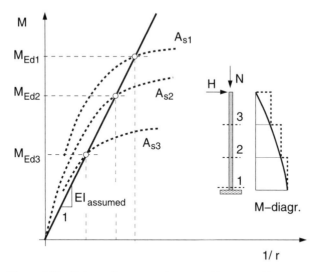

Figure 2.32. *Column design satisfying equilibrium and kinematic compatibility at multiple (3) sections*

To achieve a more economical design, a finer division is required. Such designs, requiring that both equilibrium and kinematic compatibility are satisfied at a sufficient number of cross-sections, are feasible for long compression elements, such as pylons for suspension and cable-stayed bridges.

For regular building structures, it is necessary to simplify. In such structures, the whole column length is normally provided with the same reinforcement, designed to resist the maximum moment within the column length. Compatibility is therefore, satisfied at most, at one section of the columns. Considering that a structure is checked for many load cases, and that each member is designed for the most critical load case, which may be different for different members, it is clear that the total structure may become considerable stiffer than assumed in the analysis.

For all but deformation-induced forces in statically indeterminate structures, this will represent a conservative design. Because of the stiffer structure, deformation-induced forces will increase. This should be borne in mind in the design process, so that necessary precautions may be taken for elements attracting significant forces from imposed deformations.

2.9.3. *Optimization*

A design based on assumed stiffness values may not represent an optimal design. To consider this aspect a bit further, a single, statically determinate column is considered. The column is analyzed based on different stiffness assumptions (EI), but with the same first-order moment (M_0) and axial force in all cases. Equilibrium and curvature compatibility are satisfied at the critical section only.

The column is defined by constant section and reinforcement, constant relative axial force $n = N_{Ed}/(f_{cd}A_c) = 0.4$, slenderness $L_0/h = 20$, $h'/h = 0.8$, elastoplastic reinforcement with $\varepsilon_{sd} = 0.0024$ and the conventional "parabola-rectangle" stress–strain diagram for concrete $(\varepsilon_{c2} = 0.002, \varepsilon_{cu2} = 0.0035)$.

Moment-curvature relationships for the column are given in Figure 2.33(a) in terms of relative moments $m = M/(f_{cd}A_c h)$ and section stiffnesses $\overline{EI} = EI/(f_{cd}A_c h^2)$. The external first-order moment is taken as $m_0 = 0.25$. For convenience, subscripts Ed and Rd are omitted.

For simplicity, the elastic analysis is carried out with the moment magnifier approach by $M = M_0/(1 - N/N_{cr})$, or dimensionless by:

$$m = m_0/(1 - n/n_{cr}) \quad \text{with} \quad n_{cr} = \overline{EI}/(L_0/h)^2$$

Identical results can be obtained using the additional moment formulation, $M = M_0 + M_2$, dimensionless given by:

$$m = m_0 + m_2 \quad \text{with} \quad m_2 = \frac{n}{10}\left(\frac{L_0}{h}\right)^2\frac{h}{r}$$

By varying the assumed stiffness (or alternatively, the assumed curvature), different amounts of reinforcement are required to satisfy equilibrium and curvature compatibility. Five cases are considered in Figure 2.33(a), but the secant stiffnesses are shown only for the first and last case. Reinforcement requirements for the five cases are identified in terms of the mechanical reinforcement ratios, $\omega = A_s f_{yd}/(A_c f_{cd})$, in Figure 2.33(b).

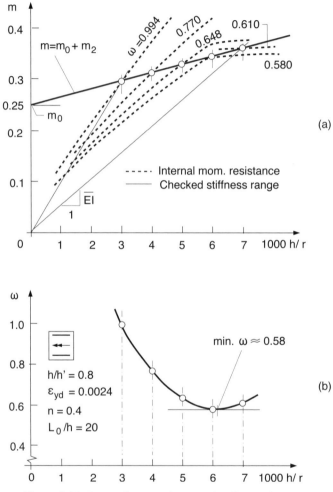

Figure 2.33. *Design for several assumed stiffness values: (a) moment-curvature diagrams, (b) reinforcement satisfying equilibrium and compatibility*

The minimum (optimal) reinforcement is obtained for $\omega \approx 0.58$. This corresponds approximately to the case with $\omega = 0.580$ in Figure 2.33(a). It can be seen that the external moment line almost becomes a tangent to the resisting moment-curvature curve for this case (at $1,000h/r = 6$). As discussed before, the tangent point corresponds to instability. The optimum design is in other words obtained, as we know from before, for the stiffness assumption and selected reinforcement that give stability failure. Design based on assumed stiffnesses that are smaller or greater than this optimal stiffness are conservative, as seen in the figure.

In practice, often involving multi-level and multi-bay frames and a number of load cases and combinations, it is most often necessary to simplify. This involves both stiffness assumptions and relaxation of compatibility requirements. Some practical adaptions are reviewed below.

2.10. Practical linear elastic analysis

2.10.1. *Stiffness assumptions*

2.10.1.1. *General*

While computed axial forces in columns of a structure are normally little influenced by flexural stiffness assumptions, bending moments (and shears) can be significantly affected. Bending moments are in some cases functions of the real, absolute stiffness values of the structural members, and in other cases functions of only the ratios between the stiffnesses values. Whether one or the other is the case is a function of the type of external action and type of analysis to be carried out. Simply stated, if the external action is applied loads (gravity, wind etc.), the structural members must be able to carry the resulting internal forces. If, on the other hand, the external action is due to environmental changes, creep, shrinkage and settlements, the structural members must be able to accommodate the resulting deformations. The latter is generally referred to as imposed deformations, and the associated forces as deformation-induced (or imposed) forces.

Some comments on stiffness assumptions for use in linear elastic frame analyses are appropriate before discussing specific methods. Both second-order and first-order analyses are considered, as well as different external actions.

2.10.1.2. *First-order analysis*

Effects on column moments of external loads (permanent and variable) due to gravity, wind, etc., are functions of the *ratios* between the stiffness of the compression members (columns) and the stiffness of the flexural (beam, plate) members. So it does not matter whether the stiffness assumption is twice, or half, the correct value, as long as the ratio is reasonably correct. EC2 (5.4) suggests the use of uncracked cross-section stiffnesses for such cases.

On the other hand, imposed deformations inflicted by temperature changes, etc., cause rotations and displacements. In statically indeterminate structures, these give rise to deformation-induced moments (forces). These moments are functions of *real* stiffness values. In some cases, such as, for instance, for columns in a multi-bay frame (bridge) with a stiff connecting beam (superstructure), deformation-induced column moments are primarily a function of the *real* stiffnesses of the columns only. Then, the moments are proportional to the assumed column stiffness.

For imposed deformation effects seen in isolation, it is worth having in mind that it is the capacity for accommodating a computed deformation (rotation, displacement), and not the capacity for "carrying" loads, that should be the important consideration. Or, in other words, it is deformation capacity, or ductility, rather than load capacity, that should be the focus in design.

Nevertheless, in todays practice, it is still computed forces that are the basis in design also for effects of imposed deformations. Use of gross, uncracked section stiffnesses, such as $EI = E_{cd}I_c$, would attract unrealistically large moments and shears. It is therefore important to assume realistic stiffnesses, and in particular realistic column stiffnesses, when computing deformation-induced moments. EC2 (5.4) recommends use of cracked section stiffness for columns in such cases.

2.10.1.3. *Second-order analysis*

Moments from second-order load effects in isolated, statically determinate members are a function of representative *real* stiffness values. If the stiffness for such a member is assumed greater than its real values, computed column deflections, and therefore second-order moments, will be underestimated.

For a framed column (as part of a frame), the total load effects are functions also of the ratio between column and beam stiffnesses, and the situation is consequently more complex. With a high column to beam stiffness assumption, the column will attract larger moments at column ends, but will, on the other hand, display smaller deflections and corresponding second-order effects (moments) between the column ends. Code approaches for such problems are briefly reviewed in the sections below.

In approximate methods that make use of critical (buckling) loads for the columns, these should be computed with low, rather than high, stiffness assumptions. Effective length factors, used for computing such buckling loads, are dependent on the ratio of the rotational end restraints stiffness to the column stiffness. A low ratio assumption is conservative in that it gives rise to a larger effective length than a high assumption would have.

2.10.2. EC2 approach

A linear second-order analysis in ULS, and subsequent conventional design, may be adopted within the EC2 regulations. EC2 (5.8.2) states (in a note) that analysis based on linear material properties may be carried out with stiffness values used in the nominal stiffness method. Thus:

$$EI = K_c E_{cd} I_c + E_s I_s \qquad [2.94]$$

may be adopted for compression members. Creep effects are included in the K_c factor (see section 2.11.3).

For the flexural members, referred to as "adjacent" members in EC2 (5.8.2), the section stiffness can be computed taking into account partial cracking and tension stiffening. As a simplification, the fully cracked stiffness may be assumed. Since it is the ULS that is of interest, and considering many uncertainties, use of the fully cracked section stiffness seems most reasonable. Possible creep is to be included by using an effective concrete creep modulus. Thus:

$$EI = EI_{cracked} \quad \text{with} \quad E_{cd,eff} = E_{cd}/(1 + \phi_{ef}) \qquad [2.95]$$

may be adopted for the flexural members according to EC2.

For load effects computed by a second-order elastic analysis based on such stiffnesses, it is customary to design (proportion) member sections at the material failure state, and without any particular checking of compatibility.

As already mentioned, effective lengths are also dependent on relative stiffnesses. For the assessments of end restraints in computations of effective lengths in the simplified approaches presented in sections 2.11.3 and 2.11.4, the same stiffness assumptions given above may be used.

2.10.3. *ACI 318 approach*

A similar second-order elastic analysis approach is included in the American ACI 318 Building Code. In the Commentary to the ACI 318-1977 code, it was recommended to carry out such analyses using:

$$EI = E_c I_c (0.2 + 1.2\rho\, E_s / E_c) \quad \text{and} \quad EI = 0.5 E_c I_c \qquad [2.96]$$

where $\rho = A_s / A_c$, for column and beam members, respectively. This recommendation was based on work by MacGregor *et al.* [MAC 75].

In more recent editions of ACI 318 (e.g. [AME 11]), it is recommended to use:

$$EI = 0.7 E_c I_c \quad \text{and} \quad EI = 0.35 E_c I_c \qquad [2.97]$$

for columns and beams, respectively. Further, $0.7 E_c I_c$ is recommended for uncracked walls, $0.35 E_c I_c$ for cracked walls and $0.25 E_c I_c$ for flat plates and slabs. These values include a stiffness reduction factor (that is similar to the inverse of $\gamma_{cE} = 1.2$ in EC2, 5.8.6).

Design (proportioning) of member sections for axial forces and moments obtained from such an analysis is carried out at the material failure state. Checking of compatibility is not required.

For effective length computations (in the simplified moment-magnification approach), ACI 318-11 recommends the same values given in equation [2.97].

Earlier, such as in ACI 318-1977, it was recommended that effective length calculations were made based on end restraints assessed with either:

$$EI = 0.2 E_c I_c + E_s I_s \quad (\text{with } \beta_d = 0) \quad \text{and} \quad EI = EI_{cracked} \qquad [2.98a]$$

or

$$EI_c = E_cI_c \quad \text{and} \quad EI = 0.5E_cI_c \qquad\qquad [2.98b]$$

for columns and beams, respectively.

2.11. Simplified analysis and design methods

2.11.1. *General*

Analysis and design practice for columns in frames is, in most simplified methods, based on a three-step procedure:

1. First-order moments (M_{0Ed}) and axial forces are first calculated in a conventional first-order linear elastic frame analysis based on assumed dimensions, and sectional bending stiffnesses EI, of the various compression and flexural members;

2. Second-order moments are then calculated by simplified, approximate methods, normally using axial forces (N_{Ed}) obtained from step 1;

3. Member design is finally carried out by providing that the sections have sufficient moment and axial load capacity (M_{Rd}, N_{Rd}) to resist the sum of the first- and second-order external (applied) moments (M_{Ed}) and corresponding axial forces (N_{Ed}).

In such methods, design (step 3) of compression members will generally consist of satisfying equilibrium between the external (applied) moments (M_{Ed}) and the internal resistance (M_{Rd}) at only the critical section, and then to provide the same (symmetrical) reinforcement along the full column length. For the normal case with approximately constant axial load, constant sectional dimensions and the same reinforcement (A_s) along the length, the critical section is the section with the greater moment.

Curvature compatibility is normally not checked in simplified methods. It is tacitly assumed that the structure has sufficient ductility to accommodate redistribution of forces caused by computational lack of compatibility between external and internal curvatures.

2.11.2. *Simplified second-order analysis*

Basically, there are two types of simplified methods, depending on the manner in which simplified second-order moment effects are computed. One method is frequently labeled "moment magnification method". The other formulation is curvature dependent, and often referred to as the "additional moment method" or the "model column method" (CEB [COM 77, COM 78, COM 93]). Axial forces from first-order analysis will generally be acceptable in these methods.

In EC2:2004, these methods are labeled as:

– Method based on nominal stiffness, and

– Method based on nominal curvature.

The two methods can be illustrated with reference to Figure 2.34. A similar figure was shown before, in Figure 2.3(b), in terms of a moment-displacement ($M - a$) relationship. According to the approximate equation [2.54], the displacement and curvature can be approximated by a linear relation. Thus, the external, total moment line in Figure 2.34 is also linear.

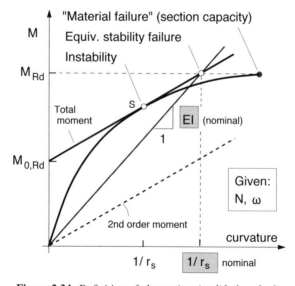

Figure 2.34. *Definition of alternative simplified methods*

The largest first-order moment that can be applied at the considered section is the moment that gives a total external moment line that becomes a tangent to the resistance curve. This is obtained at point S, which is the instability point. This maximum first-order moment that can be applied to the section may be referred to as the *first-order moment capacity* and labeled M_{0Rd}.

At instability, the total moment may be considerably smaller than the *ultimate moment capacity* M_{Rd} of the section, as illustrated in the figure. Since it is easier, and well known, how to design sections at the material failure state, for the *ultimate moment capacity* M_{Rd} of the section, it is convenient to introduce a fictitious secant stiffness that produces a second-order moment that together with M_{0Rd} becomes equal to the section capacity M_{Rd}.

The fictitious secant stiffness, denoted as "nominal" in EC2, by which this is achieved, is given by the stiffness EI in the figure. Indirectly, the design is still for stability failure. The corresponding fictitious curvature at this equivalent instability point ($1/r_{es}$ in the figure) is similarly labeled "nominal" curvature in EC2.

In the first simplified method, second-order moments in compression members are accounted for by magnifying first-order moments by a moment magnification factor (amplification factor) computed with the nominal stiffness. In the second method, second-order moment effects are accounted for by adding a nominal second-order moment corresponding to the nominal curvature shown in the figure.

These two approaches can also be identified in the axial force-moment (N–M) interaction diagram in Figure 2.35, where the dashed line is obtained with the nominal stiffness EI.

Both of these methods are well known. It is often a question of preference whether to use one or the other in some cases, while in other cases one is to be preferred to the other.

Details of both methods as specified in EC2:2004 are reviewed below. The formal derivation of the methods have previously been given and discussed in sections 2.5.1, 2.5.2 and 2.5.3.

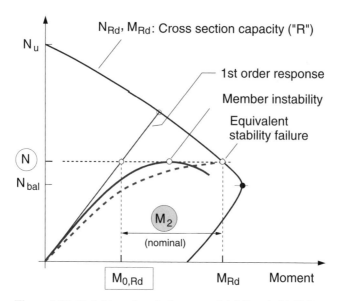

Figure 2.35. *Definition of equivalent material failure in M–N diagram*

2.11.3. *Method based on nominal stiffness*

The presentation below deviates somewhat in form, but not in content, from the EC formulations, in that moment magnification factors and a moment correction factor are introduced. In this manner, the similarity with the original derivations in sections 2.5.1 and 2.5.2 is maintained. In this book, it is distinguished for the sake of clarity between the columns with and without transverse loads between the column ends.

2.11.3.1. *Members with transverse loading at or between ends*

For compression members (columns) with transverse loading, the total design moment may be defined by:

$$M_{Ed} = B_{m0}M_{0Ed} \qquad [2.99a]$$

where

$$B_{m0} = 1 + \frac{\beta}{\dfrac{N_B}{N_{Ed}} - 1} \qquad [2.99b]$$

or, rearranged,

$$B_{m0} = \frac{1 + (\beta - 1)\frac{N_{Ed}}{N_B}}{1 - \frac{N_{Ed}}{N_B}}$$ [2.99c]

Here,

B_{m0} is a *moment magnification factor* accounting for second-order effects;

M_{0Ed} is the maximum first-order moment (including effects of imperfections);

N_{Ed} is the design value of the axial force;

N_B is the buckling (critical) load based on the nominal stiffness and the effective (buckling) length L_0 of the member;

β is a factor that depends on the moment distribution along the column axis.

For members with constant cross-section and axial load,

$$\beta = \frac{\pi^2}{c_0} \quad \text{and} \quad N_B = \frac{\pi^2 EI}{L_0^2}$$ [2.100]

The basis for the maximum moment formulation and the β and c_0 factors are presented in section 2.5.1. Typical values are summarized in Figure 2.7, and selected cases are repeated below.

Sinusoidal first-order moment: $\beta = 1$, $c_0 = \pi^2$.

Constant first-order moment: $\beta = 1.23$, $c_0 = 8$.

Triangular first-order moment: $\beta = 0.82$, $c_0 = 12$.

Parabolic first-order moment: $\beta = 1.028$, $c_0 = 9.6$.

Inverse parabolic moment: $\beta = 0.62$, $c_0 = 16$.

The design moment formulations above is suitable for members for which the maximum first- and second-order moments form at about the same section. This will typically be the case for members with transverse loading at or between ends, as seen in section 2.5.1, where the basic formulation above was derived.

Where equation [2.100] is not applicable, for instance, when maximum second-order moment and maximum first-order moment due to transverse loads do not occur at the same section, $\beta = 1$ may be adopted. Thus, for such cases:

$$B_{m0} = \frac{1}{1 - \frac{N_{Ed}}{N_B}} \qquad [2.101]$$

2.11.3.2. Members without transverse loading between ends

Columns are in the most common case part of a larger frame and subjected to axial loads and end moments. For such members, without transverse loading between ends, the maximum moment may form between the ends, as discussed previously (section 2.5.2).

In the uniform first-order case, for $M_{0Ed} = M_{01} = M_{02}$, the maximum moment can be obtained from equation [2.99] for the case with $c_0 = 8$ ($\beta = 1.23$), and written as:

$$M_{Ed} = B_{m1} M_{02} \qquad [2.102a]$$

where

$$B_{m1} = B_{m0, c_0 = 8} = \frac{1 + 0.23 \frac{N_{Ed}}{N_B}}{1 - \frac{N_{Ed}}{N_B}} \qquad [2.102b]$$

Here, N_B is the column buckling load to be calculated with the braced effective length. It may be noted that B_{m1}, in which subscript "1" signifies uniform moment (end moment ratio equal to 1), is the same factor used in section 2.4.3.

In the general case with unequal end moments at the two ends, the real first-order moment distribution is replaced by the equivalent, constant moment:

$$M_{0e} = 0.6M_{02} + 0.4M_{01} \geq 0.4M_{02} \qquad \text{[2.103]}$$

The alternative form:

$$M_{0e} = C_m M_{02} \quad \text{with} \quad C_m = 0.6 + 0.4 \frac{M_{01}}{M_{02}} \geq 0.4 \qquad \text{[2.104]}$$

will be used below. It is a convenient and familiar formulation often seen in the literature. C_m is the moment gradient modification factor (discussed in section 2.5.2). For a constant, uniform first-order moment distribution, $C_m = 1$.

The end moment ratio M_{01}/M_{02}, previously denoted by r_m (equation [2.61]), is between the numerically smallest, M_{01}, and the numerically largest, M_{02}, first-order end moments, both including imperfections. The ratio is to be entered with a positive value when the end moments give tension on the same side of the member (single curvature bending), and negative otherwise (double curvature bending). Subscripts Ed are for simplicity deleted for M_{0e}, M_{01} and M_{02}.

Then, the total design moment for the general case can be written as:

$$\begin{aligned} M_{Ed} &= B_{m1} M_{0e} = B_{m1}C_m M_{02} \\ &= B_m M_{02} \end{aligned} \qquad \text{[2.105a]}$$

where

$$B_m = B_{m1}C_m = C_m \cdot \frac{1 + 0.23 \frac{N_{Ed}}{N_B}}{1 - \frac{N_{Ed}}{N_B}} \qquad (\geq 1.0) \qquad \text{[2.105b]}$$

now a moment magnification factor is to be applied to the numerically larger first-order end moment.

It is normally acceptable according to EC2 to take the design moment according to:

$$M_{Ed} = B_m M_{02} \qquad\qquad\qquad [2.106a]$$

with

$$B_m = \frac{C_m}{1 - \frac{N_{Ed}}{N_B}} \qquad (\geq 1.0) \qquad\qquad [2.106b]$$

The implication of this simplified formulation is that the magnifier in equation [2.99c] is calculated with $c_0 = \pi^2$ ($\beta = 1$), corresponding to a sinusoidal first-order moment distribution, and combined with the same C_m as derived to correct for deviations from a first-order uniform moment distribution. This is clearly not entirely logical. Nevertheless, equation [2.106b] is the most common form in the literature, and is also the form used in the ACI 318 code (since 1971). It may give somewhat unconservative results for columns in single first-order bending, close to uniform bending as discussed in section 2.5.2.

The limit ≥ 1.0 given in parenthesis above is surprisingly not included in EC2:2004. As seen in section 2.5.2, it is not only prudent, but necessary to include this limit, and thus require that the design moment should not be taken smaller than the larger first-order moment M_{02}.

2.11.3.3. First-order moments modified by second-order sway effects

The end moments M_{01} and M_{02} in the expressions above are in EC2 defined as first-order end moments. In the braced frame case, these moments will be due to gravity loads plus possible deformation-induced forces. For this case, the moment definition is adequate.

In an unbraced or partially braced frame with side-way, the first-order definition of M_{01} and M_{02} will also include first-order effects of lateral loads (wind, etc.). For this case, the EC2 moment definition is not adequate.

As discussed in section 2.4.3 and defined by equation [2.39], the first-order end moments should in this general case be replaced by the second-order sway-modified end moments, M_{01}^* and M_{02}^*:

$$M_{0i}^* = (M_{0b} + B_s M_{0s})_i \qquad [2.107]$$

For columns in a braced frame, the side-way magnifier is clearly $B_s = 1$. Thus, M_{01}^* and M_{02}^* are identical to the first-order moments M_{01} and M_{02} for columns in such frames.

In unbraced frames, bracing columns (that contribute to the lateral resistance) will be bent into double curvature, thus giving C_m values less than 1. Because the axial force, N_{Ed}, in such a column is significantly smaller than the braced buckling load N_B, the ratio N_{Ed}/N_B will become very small. For the combination of these two effects, the lower limit of 1.0 on B_m will come into play in such cases, thus giving $B_m = 1$. This implies that the maximum moment is at a column end.

For columns that do not contribute to the lateral resistance, but instead are effectively braced by the system, local second-order effects may be significant (high N_{Ed}/N_B ratios). For such columns, behaving like fully braced columns, B_m values may become greater than 1.0.

Several codes, including ACI 318-11, give separate provisions for braced and unbraced frames, sometimes with the help of approximate criteria for deciding when a partly braced frame can be considered braced or unbraced for the purpose of column design. In the approach outlined above, involving M_{01}^* and M_{02}^*, there is no need for making such a distinction.

2.11.3.4. *Nominal stiffness*

The nominal section stiffness of a slender compression member can be estimated from:

$$EI = K_c E_{cd} I_c + K_s E_s I_s \qquad [2.108]$$

For a symmetrically reinforced section, typical for columns, it can be rewritten into the following convenient form:

$$EI = E_{cd}I_c\left(K_c + \frac{E_s}{E_{cd}}\left(\frac{i_s}{i}\right)^2\frac{A_s}{A_c}\right) \qquad [2.109]$$

1) For $\rho = A_s/A_c \geq 0.002$:

$$\left.\begin{array}{l} K_s = 1, \qquad K_c = \dfrac{k_1\,k_2}{(1+\phi_{ef})} \\[4mm] k_1 = \sqrt{\dfrac{f_{ck}\,(MPa)}{20}}, \qquad k_2 = n\,\dfrac{\lambda}{170} \leq 0.20 \end{array}\right\} \qquad [2.110]$$

If the slenderness $\lambda\,(=\ L_0/i)$ is not known, it is acceptable to take $k_2 = 0.30n \leq 0.20$.

2) For $\rho = A_s/A_c \geq 0.01$ (i.e. normally always), EI may simply be estimated with:

$$K_s = 0, \qquad K_c = \frac{0.3}{1+0.5\phi_{ef}} \qquad [2.111]$$

2.11.4. *Method based on nominal curvature*

2.11.4.1. *Members with transverse loading at or between ends*

The theoretical basis for the curvature-based method has been derived and discussed in section 2.5.3.2.

The total design moment is defined by:

$$\begin{aligned} M_{Ed} &= M_{0Ed} + M_2 \\ &= N_{Ed}(e_0 + e_2) \end{aligned} \qquad [2.112]$$

where

M_{0Ed} and N_{Ed} are the first-order moment and axial force, respectively, as defined previously;

M_2 is the nominal second-order moment;

$e_0 \; (=M_{0Ed}/N_{Ed})$ is the first-order load eccentricity and

$e_2 \; (=M_2/N_{Ed})$ is the second-order load eccentricity. Note: EC2:2004 denotes e_2 as a "deflection". In some cases, e_2 may be a physical deflection, but in other cases it is not a well-defined deflection. Therefore, load eccentricity is a better, and more general term.

The corresponding nominal second-order moment and nominal curvature (at equivalent stability failure) are illustrated in Figure 2.36.

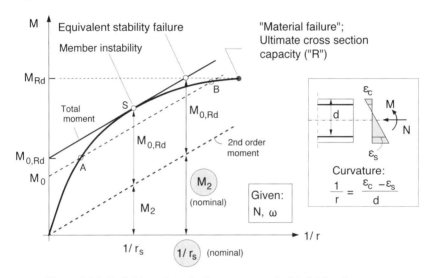

Figure 2.36. *Definition of nominal curvature method (additional moment method)*

For the normal case with constant section and axial load, the critical section will be the section with the greatest moment sum. In line with this, EC2 suggests adding the first-order and second-order moment distributions to obtain the maximum M_{Ed}, such as indicated in Figures 2.4 and 2.5. The second-order distributions may be assumed to be parabolic or sinusoidal over the effective length L_0.

Clearly, this is a possible approach provided that the formulas for inflection points are available. Such formulas are given in conjunction with the effective length and inflection points defined in Chapter 3, section 3.1.3. The approach of adding first- and second-order distributions is rather cumbersome, however.

Fortunately, for unbraced members with sway, and braced members with transverse loads, maximum first-order and second-order moments will often occur at approximately the same section. Taking M_{Ed} equal to the sum of the maximum first-order and the maximum second-order moment will be on the safe side, and normally acceptable for such cases. For members without transverse loading between the ends (Figure 2.5), a more convenient EC2 approach is given below.

2.11.4.2. Members without transverse loading between ends

For members without transverse loads between the ends, the design moment is defined by:

$$M_{Ed} = M_{0e} + M_2 \qquad (\geq M_{02})$$ [2.113]

where M_{0e} is the equivalent moment defined previously by equation [2.104].

The limit given in parenthesis above is not included in EC2:2004. As also mentioned in conjunction with the nominal stiffness method above (equations [2.105b] and [2.106b]), it is not only prudent, but necessary to include this limit, and thus require that the design moment should not be taken smaller than the larger first-order moment M_{02}.

2.11.4.3. Nominal second-order moment

The nominal (fictitious) second-order moment is defined by:

$$M_2 = N_{Ed}\, e_2 \qquad \text{with} \qquad e_2 = \frac{L_0^2}{c} \cdot \frac{1}{r}$$ [2.114]

where $1/r$ is the total curvature at the critical section.

The factor c is dependent on the curvature distribution from total moments (first- plus second order) along the member. Examples of c values are given in Figure 2.37. It should be noted that this factor is different from the c_0 factor in the moment magnifier (nominal stiffness method), as the latter is a function of the curvature from the first-order moment distribution only.

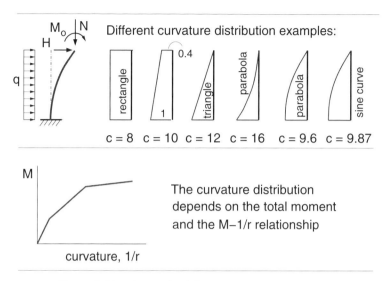

Figure 2.37. *c-factors for different curvature distributions*

It is complicated to calculate the total curvature distribution for a reinforced concrete member. Conventionally, $c = 10 \, (\approx \pi^2)$ has been used [COM 77 COM 78, COM 93]. For members without transverse loading, equation [2.56] gives an indication of the axial load effect on this factor.

2.11.4.4. *Nominal (fictitious) curvature*

The nominal curvature is defined by:

$$\frac{1}{r} = K_r \cdot K_\phi \cdot \frac{1}{r_0} \tag{2.115a}$$

where

$$\left.\begin{aligned}
\frac{1}{r_0} &= \frac{\varepsilon_{yd}}{0.45d} \qquad \text{with} \quad d = (h/2) + i_s \\
K_r &= \frac{n_u - n}{n_u - n_{bal}} \leq 1 \qquad (n_u = 1 + \omega \, ; \, n_{bal} \approx 0.4) \\
K_\phi &= 1 + \beta\phi_{ef} \qquad \text{with} \quad \beta = 0.35 + \frac{f_{ck}}{200} - \frac{\lambda}{150}
\end{aligned}\right\} \tag{2.115b}$$

Here, $1/r_0$ is the nominal "basic" curvature value, K_r reflects the effect of axial force, expressed through the axial force capacity n_u and the balanced axial load n_{bal}, and K_ϕ reflects the effect of creep. The d factor is the depth from the outer compression fiber to the tensile reinforcement. In order to cover the cases with several rebar layers and cases with rebars distributed around the circumference, etc., d is conveniently defined as an effective depth equal to half the section depth $(h/2)$ plus the radius of gyration (i_s) of the reinforcement area.

The K_r factor is illustrated in Figure 2.38, and compared to the simpler factor $K_r = 0.5/n \leq 1$, previously adopted in some manuals and codes [COM 77, NOR 89].

2.12. ULS design

2.12.1. *Simplified design methods*

The standard application of the simplified second-order analysis methods reviewed above is for columns with constant cross-section and constant, symmetrical reinforcement along the column length. A final cross-section design is therefore in such cases limited to the critical section, which for a constant axial load along the column is the section with the maximum moment.

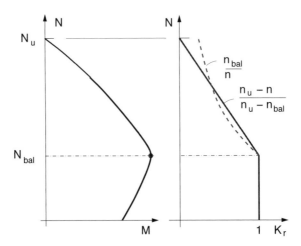

Figure 2.38. *Effect of axial force on nominal curvature*

At this section, with cross-sectional dimensions assumed in the analysis, design requires the selection of amount and placing of the reinforcement. If the section dimensions and reinforcement becomes significantly different from that assumed earlier, in the first-order analysis and in the simplified second-order moment (eccentricity) computations, a new iteration may be required.

For assumed cross-sectional dimensions, reinforcement is then selected to ensure that the section has sufficient axial force and sectional moment capacity, N_{Rd} and M_{Rd}, respectively, to carry the external design force and moment, N_{Ed} and M_{Ed}. For a given axial load, $N_{Ed}\,(=N_{Rd})$, this requires that:

$$M_{Ed} \leq M_{Rd} \qquad\qquad [2.116]$$

No matter whatever the size of computed results, the total load eccentricity (from first-order, including imperfections, and second-order effects), $e_{Ed} = M_{Ed}/N_{Ed}$, should not in any case be taken less than:

$$e_{min} = \max\,(20,\ h/30) \qquad\qquad [2.117]$$

where h is the section height in the direction of the eccentricity (EC2, 6.1).

Calculation of ultimate section capacities M_{Rd} (the maximum moment a section can resist at "material failure") for given axial forces is discussed in Chapter 1 where simplified approaches, including design aids in the form of formulas and diagrams, including $N - M$ interaction diagrams, are also discussed.

A number of computer tools, available for such calculations and indispensable in practical, day to day design work, are not discussed here.

2.12.2. *Alternative design methods*

2.12.2.1. *Design with nominal stiffness or curvature*

In the simplified methods for slender members reviewed above, the second-order moments were calculated at fictitious stability failure by using a nominal moment or a nominal curvature. This allows the design to be carried out at the material failure state. It is not required by EC2 to carry out compatibility checks.

2.12.2.2. *Direct design for first-order moment*

An alternative option to the approach above is, in the case of the curvature-based approach, to design directly for stability failure (Figure 2.36 at "S"). This normally requires computer tools, or tabulated results such as given in the CEB-FIP Buckling Manual [COM 77], where first-order moment capacities, M_{0Rd}, for given non-dimensional axial forces (n_{Ed}) are tabulated for various slenderness (L_0/h) and mechanical reinforcement (ω) ratios. Then, for a given $n(= N_{Ed}/(f_{cd}A_c)$ and L_0/h, the necessary reinforcement ω must be chosen such that:

$$M_{0Ed} \leq M_{0Rd}(n, L_0/h) \qquad [2.118]$$

where M_{0Ed} is the maximum first-order moment along the member, or, in the case of end-moment loaded members without transverse loads:

$$M_{0Ed} = C_m M_{0e} \geq M_{02} \qquad [2.119]$$

2.12.2.3. *Design with assumed curvature or stiffness*

The procedures follow those of the simplified EC2 methods, except that the nominal curvature or stiffness is to be taken equal to assumed values rather than the given nominal values. Such approaches are discussed in section 2.12.1.

For a given axial load, $N_{Ed} (= N_{Rd})$, section design requires that the total moment satisfies

$$M_{Ed} \leq M_{Rd,1/r} \qquad [2.120]$$

This is similar to equation [2.116], except that the moment capacity $M_{Rd,1/r}$ now is the moment resistance at the assumed curvature $1/r$ rather than the ultimate moment capacity M_{Rd}.

In cases when the stiffness is assumed, the curvature to design for is given by $1/r = M_{Ed}/EI$, where EI is the assumed stiffness.

When computational tools, or tabulated moment-curvature relationships, are available for given sections, reinforcement and axial forces, such approaches are convenient in some cases. Otherwise, they do not lend themselves to day-to-day design work. An advantage of such approaches is that they involve some elements of compatibility checks.

2.12.3. *Design example – framed column*

2.12.3.1. *Frame and member information*

In the design of columns in multi-level, multi-bay frames, several load combinations must be considered. For instance, the combination that produces the greatest axial load or the greatest moment in the column. In the latter case, loads may be combined such as to produce the greatest moment in single curvature bending, or the greatest moment in double curvature bending. The latter case will generally give greater first-order moments than the former, while for second-order moments, it is the other way around.

A portion of the a braced frame is shown in Figure 2.39(a). The middle column (A–B), subjected to a loading providing the most unfavorable uniform bending, is considered. Details of the first-order analysis are not shown.

Material data:

C35/45, $\gamma_c = 1.5$; B500, $\gamma_s = 1.15$, $E_s = 200$ GPa;

Concrete: $f_{cd} = \alpha_{cd} f_{ck}/\gamma_c = 0.85 \cdot 35/1.5 = 19.8$ MPa;

Reinforcement: $f_{yd} = f_{yk}/\gamma_s = 500/1.15 = 435$ MPa, $\varepsilon_{yd} = f_{yd}/E_s = 0.00217$.

Column data:

Rectangular section; assume $b/h = 350/350$ mm, $h'/h = 0.8$.

Mechanical reinforcement ratio: $\omega = \dfrac{A_s f_{yd}}{A_c f_{cd}} = \dfrac{435 A_s}{19.8 A_c} = 21.97 \dfrac{A_s}{A_c}$.

Assume 2% reinforcement: $A_s = 0.02 A_c$: $\omega = 21.97 \cdot 0.02 = \underline{0.439}$

Braced effective length: $L_0 = 4$ m, calculated with end restraints based on $EI = E_c I_c$ for the column and $EI = 0.5 E_c I_c$ for the beams.

Mechanical slenderness: $\lambda = L_0/i = 4.0/(0.29 \cdot 0.35) = 39.4$

Local member imperfections: $e_i = L_0/400 = 0.010$ m

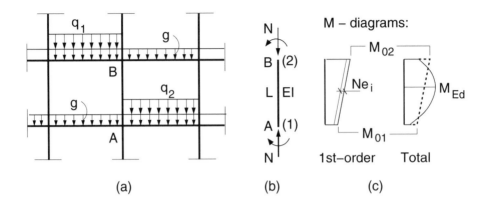

Figure 2.39. *Design example: (a) frame with loading, (b) considered column and (c) first-order and total moment diagrams*

First-order load effects:

M_1 – first-order moment without imperfection effects;

$M_0 = M_1 + N_{Ed}\, e_i$ – first-order moment including imperfection effects.

From first-order frame analysis:

Permanent actions: $N_G = 870$ kN; $M_{1A} = 16$ kNm; $M_{1B} = 18$ kNm;

Variable actions: $N_Q = 530$ kN; $M_{1A} = 60$ kNm; $M_{1B} = 70$ kNm.

Thus, end 2 (the end with the largest moment sum) will be at end B.

Load combination: Load factors: $\gamma_G = 1.35$ and $\gamma_Q = 1.5$.

$N_{Ed} = 1.35 \cdot 870 + 1.5 \cdot 530 = 1{,}970$ kN;

$M_{02} = \max M_{0Ed} = 1.35 \cdot 18 + 1.5 \cdot 70 + 1{,}970 \cdot 0.010 = 149$ kNm;

$M_{01} = \min M_{0Ed} = 1.35 \cdot 16 + 1.5 \cdot 60 + 1{,}970 \cdot 0.010 = 131$ kNm.

Creep:

$t_0 = 28$ days; RH $= 50\%$;

$h_o = 2A_c/u = 2 \cdot 350 \cdot 350/2(350 + 350) = 175$ mm;

EC2, Figure 3.1, gives $\phi(\infty, t_0) = 2.0$.

Effective creep factor:

Quasi-permanent load: $\psi_2 = 0.6$ (long-term portion of N_Q);

$N_{qp} = N_G + \psi_2 N_Q = 870 + 0.6 \cdot 530 = 1{,}188$ kN;

At the end 2 (B): $M_{0qp} = 18 + 0.6 \cdot 70 + 1{,}188 \cdot 0.010 = 72$ kNm;

$\phi_{ef} = \phi(\infty, t_0) \cdot (M_{0qp}/M_{0Ed}) = 2.0(72/149) = \underline{0.97}$.

Slenderness limit for neglect of second-order effects:

The alternative NS-EC2 method reviewed in section 2.6.3 is used;

$n = N_{Ed}/(A_c f_{cd}) = 1.970/19.8 \cdot 0.35^2) = 0.812$;

$k_a = (i_s/i)^2 = (0.5h'/0.29h)^2 = 1.9$.

Normalized slenderness:

$$\lambda_n = \lambda\sqrt{\frac{n}{1 + 2k_a\omega}} = 39.4\sqrt{\frac{0.812}{1 + 2 \cdot 1.9 \cdot 0.439}} = 21.7;$$

Moment ratio $r_m = M_{01}/M_{02} = 131/149 = 0.88$;

Creep factor: $A = 1.25/(1 + 0.2\phi_{ef}) = 1.05 > 1.0$. Thus, $A = 1$.

Limit: $\lambda_{n,lim} = 13(2 - r_m)A = 13(2 - 0.88) \cdot 1.0 = 14.6 < \lambda_n = 21.7$

The limit is exceeded: second-order effects must be accounted for.

2.12.3.2. Design moments – method with nominal stiffness

Nominal stiffness:

1) *Alternative 1:* Equations [2.108] with [2.110] with assumed 2% reinforcement:

$$EI = E_{cd}I_c \left(K_c + \frac{E_s}{E_{cd}}\left(\frac{i_s}{i}\right)^2\frac{A_s}{A_c}\right), \quad K_c = \frac{k_1 k_2}{1 + \phi_{ef}}$$

$$k_1 = \sqrt{\frac{f_{ck}}{20}} = 1.323, \quad k_2 = n\frac{\lambda}{170} = 0.812\frac{39.4}{170} = 0.187 \quad (\leq 0.2);$$

$$K_c = 1.323 \cdot 0.187/(1 + 0.97) = 0.126;$$

$$f_{cm} = f_{ck} + 8 = 43 \text{ MPa}, E_{cm} = 22(f_{cm}/10)^{0.3} = 34.1 \text{ GPa};$$

$$E_{cd} = E_{cm}/\gamma_{cE} = 34.1/1.2 = 28.4 \text{ GPa (EC2, 5.8.6(3))};$$

$$EI = E_{cd}I_c(0.123 + \frac{200}{28.4}\, 1.9 \cdot 0.02) = \underline{0.39E_{cd}I_c}.$$

2) *Alternative 2*: Equation [2.108] with [2.111] can be used:

$$EI = K_c E_{cd}I_c = \frac{0.3}{1 + 0.5\phi_{ef}}E_{cd}I_c = \underline{0.20E_{cd}I_c};$$

We choose to proceed with the most favorable (largest) stiffness;

$$EI = 0.39\, E_{cd}I_c = 0.39 \cdot 28.4 \cdot 10^3 \cdot 0.35^4/12 = \underline{13.9} \text{ MNm}^2.$$

Nominal buckling load: $N_B = \dfrac{\pi^2\, EI}{L_0^2} = \dfrac{\pi^2 \cdot 13.9}{4.0^2} = \underline{8.57} \text{ MN}$

External design forces (N_{Ed}, M_{Ed}):

The column has no transverse loading, only end moment loading.

1) *Alternative 1*: The moment magnifier by equation [2.105b] applies:

$$N_{Ed}/N_B = 1{,}970/8{,}570 = 0.23;$$

$$C_m = 0.6 + 0.4\, r_m = 0.6 + 0.4 \cdot 0.88 = 0.96 \ (\geq 0.4);$$

$$B_m = \frac{1 + 0.23 \cdot 0.23}{1 - 0.23}\, C_m = 1.367\, C_m;$$

$$M_{Ed} = B_m\, M_{02} = (1.367 \cdot 0.96)149 = \underline{196} \text{ kNm}.$$

2) *Alternative 2*: The moment magnifier by equation [2.106b] applies.

With this simplified magnifier, a 5.0% smaller design moment prediction, $M_{Ed} = \underline{186}$ kNm, is obtained. It may be noted that it is still larger than the curvature-based prediction (of 180 kNm) based on nominal curvature below.

Check of minimum eccentricity: $e_{min} = \max\left(20\,\text{mm}, h/30\right) = 20\,\text{mm}$.
$e_{Ed} = M_{Ed}/N_{Ed} = 196/1{,}970 = 0.100 > e_{min} = 0.020$.

Reinforcement: for $n = 0.812$, $m = M/(f_{cd}A_ch) = 196 \cdot 10^{-3}/(19.8 \cdot 0.35^2 \cdot 0.35) = 0.231$, two rebar layers $h' = 0.8h$ apart ($d' = 0.1h$).

From the N–M interaction diagram [DES 05] for this case (also given in Chapter 1, Figure 1.29), a value of $\omega \approx 0.46 - 0.47$ is found, corresponding to $A_s \approx 2.10 - 2.14\%$.

This is somewhat greater than the assumed $A_s = 2\%$ ($\omega \approx 0.439$) in the stiffness and second-order moment calculation. However, recalculation with a new A_s assumption is not considered necessary as only slightly smaller moments can be expected.

2.12.3.3. *Design moments – method with nominal curvature*

Assume as before 2% reinforcement ($\omega = 0.439$).

$$\frac{1}{r_0} = \frac{\varepsilon_{yd}}{0.45d} = \frac{0.00217}{0.45(0.9 \cdot 0.35)} = 0.0153 \ (\text{m}^{-1});$$

$$K_r = \frac{n_u - n}{n_u - n_{bal}} = \frac{1.439 - 0.812}{1.439 - 0.4} = 0.604;$$

$$\beta = 0.35 + \frac{f_{ck}}{200} - \frac{\lambda}{150} = 0.35 + \frac{35}{200} - \frac{39.4}{150} = 0.262;$$

$$K_\phi = 1 + \beta\phi_{ef} = 1 + 0.262 \cdot 0.97 = 1.254;$$

Nominal curvature: $\dfrac{1}{r} = K_r \cdot K_\phi \cdot \dfrac{1}{r_0} = 0.604 \cdot 1.254 \cdot 0.0153 = \underline{0.0116}$;

Second-order eccentricity: $e_2 = \dfrac{L_0^2}{10} \cdot \dfrac{1}{r} = \dfrac{4.0^2}{10} \cdot 0.0153 = \underline{0.0186}$ m.

External design forces (N_{Ed}, M_{Ed})

$N_{Ed} = 1{,}970$ kN,

$$M_{Ed} = M_{0e} + M_2 = C_m M_{02} + M_2 = 0.96 \cdot 149 + 1970 \cdot 0.0186 = 180 \text{ kNm};$$

$$e_{Ed} = M_{Ed}/N_{Ed} = 180/1{,}970 = 0.091 \text{ m} \quad (>e_{min}).$$

The nominal curvature method gives in this case a smaller design moment than the nominal stiffness method. It is about 8% smaller than that obtained by the nominal stiffness method based on equation [2.105b] ($\beta = 1.23$), and about 3% smaller than that based on equation [2.106b] ($\beta = 1$).

Chapter 3

Approximate Analysis Methods

3.1. Effective lengths

3.1.1. *Definitions and exact member analysis*

3.1.1.1. *General*

Flexural critical loads and corresponding effective lengths of elastic compression members are important parameters in the assessment of stability of framed structures, and also for estimating second-order load effects by approximate methods. Most national and international design codes describe such approaches applied to framed structures with both linear and nonlinear material properties. Simplified system instability analysis by methods involving restrained compression members considered in isolation is allowed by most codes and is common in practical design work. This is so despite the increased availability and capacity of computational methods. It is an advantage to have alternative methods, and it is believed that there will be continued use of simplified methods that allow mental calculation and that may be used instead of, or in parallel with, computational approaches.

A compression member's effective length can formally be determined from the critical load N_{cr} (N_B in EC2) given by:

$$N_{cr} = \frac{\pi^2 EI}{L_0} \qquad [3.1]$$

as

$$L_0 = \beta_0 L = \sqrt{\frac{\pi^2 EI}{N_{cr}}} \qquad\qquad [3.2]$$

where

$$\beta_0 = L_0/L \qquad\qquad [3.3]$$

is the effective (buckling) length factor. A brief review of elastic critical load and effective length, presented from another angle, is given in section 2.3.1 of Chapter 2.

For linear elastic members, the section stiffness EI will be $EI = E \cdot I$. Applied to members with nonlinear materials, EI will represent a secant stiffness, as discussed in Chapter 2.

3.1.1.2. Isolated member analysis and restraints

The interaction between a column and the remainder of the structure, of which it is a part, can be reflected by spring restraints as illustrated for member AB in Figure 3.1. If the lateral spring (bracing) is very stiff ($k_L \approx \infty$), the member is *fully braced*. If it is zero ($k_L = 0$), the member is *unbraced* (free-to-sway). These are the two cases considered here. By use of results from these two cases, more complex systems can be treated, as shall be seen later in this chapter.

The rotational springs have stiffnesses defined by:

$$k_{\theta j} = (k_c + k_b)_j \qquad j = A, B \qquad\qquad [3.4]$$

where, at end A, k_c is the rotational stiffness of the column above joint A and k_b of the two beams framing into joint A. They are again dependent on the rest of the frame through their far end boundary conditions. This is the same for joint B. For columns that do not interact with others, the end restraints will always be positive. Normally, this will also be so for framed columns, but not in cases when, for instance, a column above or below contribute with a significant negative k_c contribution to the stiffness sum (equation [3.4]).

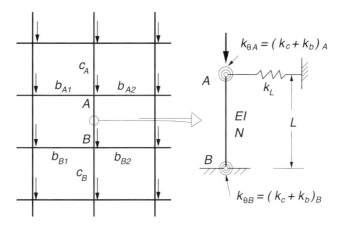

Figure 3.1. *Isolated member analysis*

The system stiffness matrix of a member with length L, section stiffness EI, axial force N and rotational and lateral restraints as defined in Figure 3.1, can be written as:

$$\begin{bmatrix} R_A \\ R_B \\ R_L \end{bmatrix} = \frac{EI}{L} \begin{bmatrix} C + \kappa_{\theta A} & S & -\frac{C+S}{L} \\ S & C + \kappa_{\theta B} & -\frac{C+S}{L} \\ -\frac{C+S}{L} & -\frac{C+S}{L} & \frac{2(C+S)-(pL)^2}{L^2} + \frac{k_L}{EI/L} \end{bmatrix} \begin{bmatrix} \theta_A \\ \theta_B \\ \Delta \end{bmatrix} \quad [3.5]$$

where

$$\kappa_{\theta j} = \frac{k_{\theta j}}{(EI/L)} \qquad j = A, B \qquad\qquad [3.6]$$

are the non-dimensional rotational restraint stiffnesses at end A and B. Further, θ_A, θ_B and Δ are the end rotations at A and B and the relative lateral translation of the two ends, respectively, and R_A, R_B and R_L are the associated joint moments and lateral load. C and S are the so-called stability functions, which may be defined by:

$$C = \frac{c}{c^2 - s^2}; \qquad S = \frac{s}{c^2 - s^2} \quad \left(= C\frac{s}{c} \right) \qquad\qquad [3.7]$$

where

$$c = \frac{1}{(pL)^2}\left[1 - \frac{pL}{\tan pL}\right]; \qquad s = \frac{1}{(pL)^2}\left[\frac{pL}{\sin pL} - 1\right] \qquad [3.8]$$

The lower-case c and s functions above are simply equal to one-third of the corresponding functions (ϕ and ψ) presented and tabulated in [TIM 61].

In first-order theory, when the axial force is zero, C and S take on the familiar values of 4 and 2. In second-order theory, they incorporate the second-order effects of axial forces on deformations. In the literature, various formulations of these, or equivalent functions, can be found. The above-mentioned formulation can be found, for instance, in [GAL 68] or [GAL 08].

3.1.1.3. *Exact effective length factor for two fundamental cases*

Exact elastic effective length factors can readily be determined from the zero determinant condition of the stiffness matrix (defined above) of the restrained compression member.

For instance, for the braced case ($\Delta = 0$), the buckling condition is given by the determinant of the sub-matrix consisting of the two first lines and rows in equation [3.5], and can be expressed by:

$$(C + \kappa_{\theta A})(C + \kappa_{\theta B}) - S^2 = 0 \qquad [3.9]$$

This may be written in a more convenient form. Substituting for the upper-case C and S from equation [3.7] and dividing by $\kappa_{\theta A}\kappa_{\theta B}$ gives:

$$\frac{1}{\kappa_{\theta A}\kappa_{\theta B}} + c\left(\frac{1}{\kappa_{\theta A}} + \frac{1}{\kappa_{\theta B}}\right) + c^2 - s^2 = 0 \qquad [3.10]$$

Introducing the definition of the lower case c and s, it can be shown that $c^2 - s^2$ can be written by:

$$c^2 - s^2 = \frac{1}{(pL)^2}\left[\frac{\tan(pL/2)}{(pL/2)} - 1\right] \qquad [3.11]$$

For N equal to the critical load $N_{cr} = \pi^2 EI/(\beta_0^2 L^2)$, pL can be replaced by $pL = \pi/\beta_0$. Then, by substituting this and $c^2 - s^2$ into equation [3.10], the following well-known transcendental equation results for the braced case:

$$\frac{(\pi/\beta_0)^2}{\kappa_{\theta A}\kappa_{\theta B}} + \left(\frac{1}{\kappa_{\theta A}} + \frac{1}{\kappa_{\theta B}}\right)\left(1 - \frac{(\pi/\beta_0)}{\tan(\pi/\beta_0)}\right) + \frac{\tan(\pi/2\beta_0)}{(\pi/2\beta_0)} = 1 \qquad [3.12]$$

Similarly, for the unbraced case, the following equally well-known transcendental equation can be derived:

$$\frac{(\pi/\beta_0)^2 - \kappa_{\theta A}\kappa_{\theta B}}{\kappa_{\theta A} + \kappa_{\theta B}} = \frac{(\pi/\beta_0)}{\tan(\pi/\beta_0)} \qquad [3.13]$$

Effective lengths for these two fundamental cases are illustrated in Figure 3.2. Physically, the effective length is equal to the distance between inflection points (located within the member length or on the mathematical continuation of the buckled member shape).

These transcendental equations are frequently given in terms of dimensionless restraint flexibility parameters [HEL 96b, GAL 68]. The most well known of such parameters is the G factor, which, with the conventional definition adopted in a number of codes (e.g., ACI 318 [AME 11], where it is denoted ψ), will always become positive. This is one of several limitations of the conventional definition [HEL 96a, HEL 96b]. With a generalization given by:

$$G_i = b_o\frac{(EI/L)}{k_{\theta j}} = \frac{b_o}{\kappa_{\theta j}} \qquad j = A, B \qquad [3.14]$$

which allows for the use of either positive or negative values, the corresponding transcendental expressions will not be subject to any limitation. The coefficient b_o is a reference restraint stiffness coefficient, normally taken equal to $b_o = 6$ (corresponding to that for a beam bent in antisymmetric curvature) for the unbraced case and $b_o = 2$ (beam bent in symmetric curvature) for the braced case. Another flexibility definition, considered below, is used by EC2:2004.

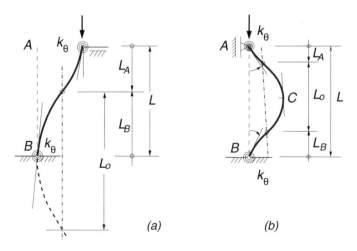

Figure 3.2. *Typical effective lengths and inflection point locations for cases with positive rotational restraints: (a) unbraced and (b) fully braced*

3.1.2. *EC2 effective length of isolated members*

Effective lengths are given as functions of restraint flexibilities. The expressions are based on curve fitting according to Westerberg [WES 04]. It is not stated in the code, but can be assumed, that the results were obtained for, and intended to be applied to, columns with constant EI and N along the length.

1) For unbraced members (Figure 3.2a):

$$\beta_0 = \max\left[\sqrt{1 + 10\,\frac{f_{\theta A}\,f_{\theta B}}{f_{\theta A} + f_{\theta B}}}\;;\;\left(1 + \frac{f_{\theta A}}{1 + f_{\theta A}}\right)\left(1 + \frac{f_{\theta B}}{1 + f_{\theta B}}\right)\right] \qquad [3.15]$$

2) For braced members (Figure 3.2b):

$$\beta_0 = 0.5\sqrt{\left(1 + \frac{f_{\theta A}}{0.45 + f_{\theta A}}\right)\left(1 + \frac{f_{\theta B}}{0.45 + f_{\theta B}}\right)} \qquad [3.16]$$

where

$$f_{\theta j} = \frac{\theta_j}{M_j} \cdot \frac{EI}{L} \quad \left(= \frac{(EI/L)}{k_{\theta j}}\right) \qquad j = A, B \qquad \text{[3.17]}$$

is the relative (dimensionless) rotational flexibility of the restraints at end $j = A$ and $j = B$. Here, $(\theta/M)_j$ is the restraint flexibility (per definition equal to the rotation divided by the moment necessary to inflict that rotation). The rotational flexibility is the inverse of the rotational stiffness, and varies between zero (fully fixed end) and infinity (hinged, or pinned, end).

If an adjacent compression member at a joint contributes to the rotation at buckling, then EI/L of the considered column should be replaced by the sum $(\sum EI/L)$ over the columns meeting at the joint:

$$f_{\theta j} = \frac{\theta_j}{M_j} \cdot \left(\sum \frac{EI}{L}\right)_j \qquad \text{[3.18]}$$

The code does not explain the basis for this specification. It can be shown, however, that it follows from a vertical interaction assumption given below by equation [3.33], and it will be discussed further there.

It should be noted that EC2 uses the letter symbol k_j for relative (dimensionless) flexibilities rather than the symbol $f_{\theta j}$ adopted above. It is rather common practice in the literature to use k for the absolute stiffness (spring stiffness, etc.). In line with this, the symbol $k_{\theta j}$ is used here for rotational stiffness and $\kappa_{\theta j}$ for relative rotational stiffness.

3.1.3. *Alternative effective length expressions*

A number of other alternative approximate effective length formulas are available in the literature. An overview is given in [HEL 12]. It is an advantage that the formulas are based on physical models rather than being based on curve fitting. A set of such formulas for unbraced and braced columns, with constant N and EI along the length, were derived from one basic physical model by Hellesland [HEL 94a, HEL 07] in terms of *rotational degree of fixity factors R* defined by:

$$R_j = \frac{k_{\theta j}}{k_{\theta j} + c\,EI/L} = \frac{1}{1 + c/\kappa_{\theta j}} \qquad j = A, B \qquad [3.19]$$

where

$$k_{\theta j} = M_j/\theta_j \qquad [3.20]$$

and

$$\kappa_{\theta j} = \frac{k_{\theta j}}{EI/L} \quad \left(= \frac{1}{f_{\theta j}}\right) \qquad [3.21]$$

are the rotational restraint stiffness and the corresponding relative, non-dimensional stiffness, respectively, at end j. The c factor evolves from the analytical derivation and becomes:

  – $c = 2.4$ for unbraced columns, and

  – $c = 4.8$ for braced columns.

Note that these c factors are not related to the c factors in Chapter 2.

Such degree of fixity factors are considered to be very suitable restraint factors for practical use. They reflect physical properties of both the end restraints and of the column itself. For positive restraints, they vary between 1.0 (fully fixed; 100% fixity) and 0 (hinged; 0% fixity). For negative restraints, they may take on values outside this range. The distance from a column end to the inflection point is directly proportional to the fixity factor at that end.

For unbraced members with $c = 2.4$, β_0 and the location of the inflection points from end j can be calculated from:

$$\beta_0 = \frac{2\sqrt{R_A + R_B - R_A R_B}}{R_A + R_B}; \qquad \frac{L_j}{L} = \frac{R_j}{R_A + R_B} \qquad [3.22]$$

For braced members with $c = 4.8$, β_0 and the location of the inflection points from end j can be calculated from:

$$\beta_0 = \frac{1}{\sqrt{(1 + R_A)(1 + R_B)}}; \qquad \frac{L_j}{L} = (1 - \beta_0) \frac{R_j}{R_A + R_B} \qquad [3.23]$$

or, alternatively and with somewhat better accuracy, with β_0 by:

$$\beta_0 = \frac{2}{2 + 1.1R_{min} + 0.9R_{max}} \qquad [3.24]$$

where R_{min} and R_{max} is the algebraically smallest and largest, respectively, of R_A and R_B (for instance, if $R_A = 0.1$ and $R_B = -0.4$, then $R_{min} = -0.4$ and $R_{max} = 0.1$).

These formulas are simple and well suited for practical applications, and more accurate than the EC2 formulas. The normal application is for columns with positive restraints (R between 0 and 1), for which the accuracy is within 0 to +2 % of exact results for the ubraced case (equation [3.22]) and within $\pm$ 1 % for the braced case (equation [3.24]). Equations [3.22] and [3.24] were adopted in the Norwegian Standard NS 3473:1998 [NOR 89].

In addition, the formulas are suitable for a range of special applications that may also include negative restraints. Examples of such applications are cases involving one joint with unknown rotational restraints. For additional details, including detailed accuracy evaluations, see [HEL 07].

The inflection point locations can sometimes be of interest, for instance, in order to sketch the approximate second-order moment distribution. Furthermore, provided R_i is computed with $c = 2$, L_j/L in equation [3.22] gives the exact location of the first-order inflection point of an unbraced, rotationally restrained column subjected to a lateral top load, and can, in this case, be used directly in determining the first-order moment distribution in such a column. It is consequently a very useful relationship.

Partially braced members: Such members are dealt with in detail in section 3.3.

3.1.4. *Columns with beam restraints*

The rotational end restraints defined explicitly by springs in Figure 3.2 are in real frames due to the interaction with beams (floor plates, etc.), and, in the case of multi-story frames, also with other columns framing into the considered column end. With reference to Figure 3.3, the restraint at end j of column c may be given by:

$$k_{\theta_j} = f_j \, k_{bj} \tag{3.25}$$

where, when restraints are provided by beams,

$$k_{bj} = \left(\sum b \frac{EI_b}{L_b} \right)_j \tag{3.26}$$

Here, k_{bj} is given by the sum of the rotational restraint stiffness at j of all the beams framing into the considered end, b is the bending stiffness coefficient of the beam at the considered end, and, in the case of multi-story frames, f_j is the fraction (or multiple) of the beam restraint that is "allocated" to, or "demanded" by, the column end considered.

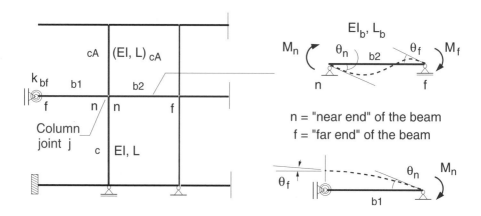

Figure 3.3. *Rotational stiffness of beams*

With the "conventional" f_j factor (equation [3.33]), the dimensionless restraint stiffness κ_θ and flexibility f_θ are given by:

$$\kappa_{\theta j} = \frac{k_{\theta j}}{(EI/L)} = \frac{1}{f_{\theta j}} = \frac{(\sum b\,EI_b/L_b)_j}{(\sum EI/L)_j} \qquad j = A, B \qquad [3.27]$$

With this definition, the R values can be expressed by:

$$R_j = 1\Big/\left(1 + \frac{c\sum EI/L}{\sum bEI_b/L_b}\right)_j \qquad j = A, B \qquad [3.28]$$

The summation in the nominator includes all columns that (at buckling) participate with the rotation of the considered joint. This is in line with common practice (also in EC2), but is not particularly accurate [HEL 96a] as discussed in section 3.2.

Above, k_{bj} reflects the horizontal interaction of the column with beams and columns in neighboring bays and f_j reflects the vertical interaction with adjacent columns and beams on other floors. Restraint mechanics are clearly complicated and are considered in more detail in the following.

3.1.4.1. *Rotational stiffness of beams – horizontal interaction*

The rotational stiffness k_b at the end n of a beam with constant sectional bending stiffness EI_b along the member length L_b can be expressed by

$$k_b = \frac{M_n}{\theta_n} = b\,\frac{EI_b}{L_b} \qquad [3.29]$$

where b is a rotational bending coefficient that can be determined from the differential equation, or more readily from the stiffness or flexibility relationship of the beam elements.

For element $b2$ in Figure 3.3, with no relative lateral displacement of the supports, the first-order stiffness coefficient can be expressed by either one of the two rotational relationships below:

$$b = \frac{6}{2 - (M_f/M_n)} \qquad \text{and} \qquad b = 4\left(1 + 0.5\,\frac{\theta_f}{\theta_n}\right) \qquad [3.30]$$

where M_n and M_f are the near end and far end beam moments and θ_n and θ_f are the corresponding rotations (Figure 3.3). The moment ratio is positive when the moments act in the same direction (clockwise or anticlockwise), and similarly for the rotation ratio.

Typical values of the rotational bending coefficient are $b = 4$ to $b = 3$ for beams with far ends clamped ($\theta_f = 0$) or pinned ($M_f = 0$), respectively. For beams hinged to the considered joint, $b = 0$ ($M_n = 0$). The same value should probably be used for beams connected to the considered joint by a flexible connection, if more accurate values cannot be documented.

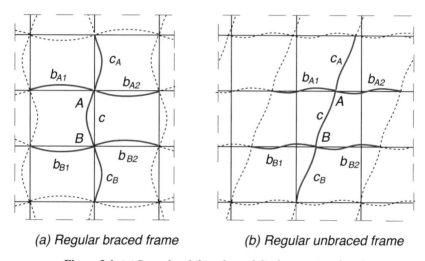

(a) Regular braced frame   (b) Regular unbraced frame

Figure 3.4. *(a) Braced and (b) unbraced displacement modes of regular frames*

For beams bent into symmetric single curvature, $b = 2$ ($\theta_f = -\theta_n$), and for antisymmetric double curvature, $b = 6$ ($\theta_f = \theta_n$). These cases are illustrated in Figure 3.4.

For element $b1$ in Figure 3.3, the left ("far end") support can translate relative to the other support ("near end"). The first-order stiffness coefficient for such an element can conveniently be expressed by:

$$b = \frac{1}{1 + \dfrac{(EI_b/L_b)}{k_{\theta f}}}$$

[3.31]

When the far end is nearly or fully fixed against rotations ($k_{\theta f} \approx \infty$), $b = 1$. If it is pinned ($k_{\theta f} = 0$), $b = 0$. If, at the far end, it is rigidly connected to an element with the same (EI_b/L_b) value as the considered one, b (equation [3.31]) will vary between 0.75 and 0.8, corresponding to $k_{\theta f}$ values of 3–4 EI_b/L_b.

The given stiffness expressions will not be significantly affected by small axial force levels (small N/N_{cr} ratios). The expressions presume rigid beam-column connections (i.e. that the angle between the members remains unchanged during the joint rotation). If the connection is not rigid, but semi-rigid (flexible), the bending coefficient b must be determined for actual degree of rigidity. See, for instance, [XU 02, HEL 09a] and others.

3.1.4.2. *Vertical interaction*

The total rotational restraint offered by restraining beams, and possibly by other restraining elements, at a joint, will have to restrain all compression elements that frame into the joint. This "vertical interaction" between a column and columns in adjacent stories can be reflected through the factor f_j that represents the "fraction" (or multiple) of the total restraint that is "allocated" to, or "demanded" by, the column end considered.

This restraint demand factor can formally be defined by:

$$f_j = \frac{M_j}{(\sum M)_j} \qquad\qquad [3.32]$$

where M_j is the moment at the considered column end j and the summation is over this moment and the corresponding moment in the adjacent column (cA in Figure 3.3). Moments are defined positive when acting in the same direction (clockwise or anticlockwise). The restraint offered at a joint will in other words be distributed to the columns framing into the joint in proportion to their moments at the joint. In single curvature regions, typical for multi-story frames with stiff columns and flexible beams, f_j will become negative for one of the columns framing into a joint, thereby indicating that also the stiffer column at the joint contributes, in addition to the beams, to the restraint of the most flexible column ([HEL 09a, HEL 09b]).

This restraint distribution, equation [3.32], has been given before for braced frames [LAI 83b], but it is valid for any frame [HEL 09a].

Equation [3.32] is not particularly useful, except for in the case of frames with sway for which the first-order lateral deflection mode due to lateral loading is similar to the lateral buckling mode. In such cases, the approximate "fraction" f_j factors can be computed by equation [3.32] based on first-order moment results (M_0).

A common, and rather "conventional", assumption is that the beam restraints are "shared" between columns of different stories meeting at a joint (j) in proportion to their EI/L values, such that:

$$f_j = \frac{EI/L}{(\sum EI/L)_j} \qquad\qquad [3.33]$$

where EI/L in the numerator is for the column considered, and $(\sum EI/L)_j$ is the sum over this (c) and the adjacent column (cA). This distribution is inherent in the "conventional" G factor definition and in the EC2 restraint formulation (equation [3.18]).

This factor, and other similar factors that also include effect of axial forces, is discussed in [HEL 96a] in conjunction with system critical load analysis, and there labeled restraint demand factor. Contrary to what has been found in stability analyses, this factor provides the stiffer column (in terms of EI/L value) with the larger restraint portion. This is normally conservative in that less restraint is provided to the critical columns than should have been. Another major deficiency is that it does not allow negative values, and thereby implies that beams provide all restraints at a joint. This is not always the case, and, in particular, not so in regions of sway frames with single curvature bending, where the major restraint at a joint is provided by the stiffer column [HEL 09b].

However, despite its deficiencies, this factor is extensively used. Results obtained, particularly in critical system load contexts, may be improved using the "method of means (MOM)" described in section 3.2.

3.1.4.3. *Restraints in regular frames – horizontal interaction*

For regular frames, it is customary to assume deflected shapes as those shown in Figure 3.4 for braced and unbraced frames, respectively.

In such cases, we can assume that the beams bend into symmetric single curvature with $b = 2$ for braced frames and into antisymmetric double curvature with $b = 6$ for unbraced frames.

3.2. Method of means

3.2.1. *General*

Hellesland and Bjorhovde's method of means [HEL 96b], denoted MOM for short, offers a simple approach for computation of effective length factors for continuous columns and frames. It makes use of isolated effective length factors obtained by the "conventional" approach (or other standard methods). The method satisfies basic system instability principles, and does, with appropriate constraints imposed, provide predictions that generally will be in excellent agreement with the results of exact system instability analyses. The presentation here largely follows that made in a recent review of the method [HEL 12].

The method should be applied to compression members that interact with each other, either directly or through other members. Compression members that do not interact with the others, such as pinned-end members in braced frames, are analyzed independently and proportioned such that local ("braced") buckling of these do not induce premature system instability.

For larger multi-story frameworks, it has been suggested [HEL 96b] that a partial application to a limited number of interacting columns in a limited region of a frame may be not only adequate, but also preferable in order not to "suppress" localized failure in a region that may be significantly more flexible than the rest of the structure. For braced structures, a partial application to a limited number of *interconnected* compression members is probably most practical and appropriate. For multi-story sway frames, this may also apply to a limited number of stories (see section 3.3.6.3 and 3.3.6.4).

3.2.2. *Method of means – typical steps*

Step 1: Determine isolated "conventional effective lengths" of the compression members based on the conventional approach,

$$\beta_{01}, \ \beta_{02}, \ \beta_{03}, \ \ldots, \beta_{0n}$$

and establish corresponding member *stability indices* defined by:

$$\alpha_i = \frac{N_i}{N_{cr\,i}} = (\beta_0^2 \, \alpha_E)_i \quad \text{where} \quad \alpha_{Ei} = \frac{N_i}{N_{Ei}} \qquad [3.34]$$

for $i = 1, 2, ...n$. N_E is, as before, the Euler buckling load of a pinned-end column.

Step 2: For larger frameworks, localize the "critical region(s)" where system instability will be initiated. This is typically that (those) surrounding the member(s) with the larger isolated member stability index (indices) from equation [3.34]. For small frames, the "critical region" may encompass the whole frame.

The member stability indices will, as mentioned before, be equal if the β_0 factors (in step 1) were the exact factors. In approximate methods, this will not be the case, however. The mean value of the member stability indices for n interacting compression members in the critical region will represent a good approximation provided that suitable lower limit constraints are imposed.

Step 3a: An improved "mean"-based effective length factor for an arbitrary member j can now be obtained from equation [3.34](or equation [2.6]) as:

$$\bar{\beta}_{0j} = \sqrt{\frac{\overline{\alpha}}{\alpha_{Ej}}} \qquad [3.35]$$

where

$$\alpha_{system} \approx \overline{\alpha} = \frac{1}{n}\left[\alpha_1 + \alpha_2 + \cdots + \alpha_n\right] \geq \lim \alpha_k \qquad [3.36]$$

$$\lim \alpha_k = \max\left[\lim \alpha_1, \lim \alpha_2, ... \lim \alpha_n\right] \qquad [3.37]$$

$$\text{where } \lim \alpha_i = \alpha_{Ei} \left(\lim \bar{\beta}_{0i}\right)^2 \quad i = 1, 2, ...n \qquad [3.38]$$

The contraints (i.e. the limits $\lim \alpha$ and $\lim \bar{\beta}_0$) are defined in step 4.

Step 3b: Alternatively, by substituting equation [3.34] into equation [3.35], $\bar{\beta}_{0j}$ can be expressed by:

$$\bar{\beta}_{0j} = \sqrt{\frac{\bar{\alpha}}{\alpha_{Ej}}} = \left[\frac{1}{n}\sum_{i=1}^{n}\beta_{0i}^{2}\,Q_{ij}\right]^{1/2} \geq \lim \bar{\beta}_{0j} \qquad [3.39]$$

where

$$Q_{ij} = \frac{\alpha_{Ei}}{\alpha_{Ej}} \left(= \frac{N_i}{N_j}\cdot\frac{L_i^2}{L_j^2}\cdot\frac{EI_j}{EI_i} = \frac{(NL)_i}{(NL)_j}\cdot\frac{(EI/L)_j}{(EI/L)_i}\right) \qquad [3.40]$$

$$\lim \bar{\beta}_{0j} = \max\left[(\lim \bar{\beta}_{0i})\sqrt{Q_{ij}}\right] \qquad i = 1, 2, ...n \qquad [3.41]$$

The expression in terms of the quotient Q_{ij} between the nominal load indices of member i and j, is often a convenient form, that may reduce computational work. Once $\bar{\beta}_{0j}$ is determined for member j, the effective length factor in any member i can be expressed in terms of this $\bar{\beta}_{0j}$ as given by:

$$\bar{\beta}_{0i} = \bar{\beta}_{0j}\sqrt{Q_{ji}} = \frac{\bar{\beta}_{0j}}{\sqrt{Q_{ij}}} \qquad i = 1, 2, ...n \qquad [3.42]$$

Step 4: Lower limit constraints: If members with relatively low nominal flexibilities (α_E), which it would have been more appropriate to classify as flexural members, have been included in the summation in equation [3.36] or [3.39], effective length predictions below theoretical minimum values may result. The approximate lower limit constraints in steps 3a $(\lim \alpha)$ and 3b $(\lim \bar{\beta}_0)$ are included to deal with such cases.

A lower effective length factor limit $(\lim \bar{\beta}_{0i})$ for a column i can be taken as *the effective length factor obtained if all members attached to its ends (including the adjacent columns) are considered to be flexural ("beam") members with first-order rotational stiffnesses that reflect these members' far end boundary conditions* [HEL 97, HEL 98]:

— *Braced case:* Restraint stiffness coefficients may for simplicity be taken as $b = 3$ for all rigidly attached beams and adjacent columns (implying negligible far end restraint), or more accurately, as b values given by equation [3.30].

— *Unbraced case:* Restraint stiffness coefficients are $b = 1$, 0 and, for simplicity, about 0.7 when the far end is fully fixed, pinned or continuous (partially restrained), respectively. Alternatively, equation [3.31] can be used.

3.2.3. *Application of the method of means*

Results using the MOM have been compared to a range of exact solutions for a number of braced and unbraced applications [HEL 96b], including unbraced multi-story, multi-bay frames [HEL 98]. An application of the MOM to stories in multi-story building structures is discussed in section 3.3.6.

Two other examples are presented below to illustrate the method in more detail. The results labeled MOM in Figures 3.5 and 3.7, for the frame in Figure 3.6, are computed by the MOM.

The MOM results in the figures are based on initial, isolated factors (step 1) that are computed using exact theory with conventional restraints. Exact solutions are obtained employing member stiffness formulations in terms of standard stability function. Axial deformations are neglected and the axial force distribution in the members is prescribed.

The agreement with the exact results in the figures must be considered very good. Branches labeled "without lower limits" show predictions that result if the lower limits in step 4 had not been applied. It can be seen that the lower limits, as defined, provide good constraints.

a) Application to two-story unbraced frame

The sample calculation will be made for the frame shown by the inserts in Figure 3.5(b) for:

$$EI_2 = 2EI_1, EI_b = EI_1, L_1 = L_2 = L_b \text{ and } N_1 = N_2.$$

Thus, $Q_{12} = 2$ and $Q_{21} = 0.5$.

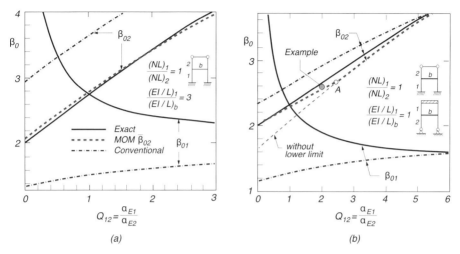

Figure 3.5. *Comparison of MOM and exact effective length factors for two-story frame with (a) flexible and (b) intermediately stiff middle beam (from [HEL 12]*

The frame will buckle sideways into an antisymmetric mode (with $b = 6$ for the middle beam), and it is therefore sufficient to consider two interconnected columns only ($n = 2$). The approximate unbraced effective length factor expression in equation [3.22] will be used to calculate isolated β factors in the sample calculation.

Step 4: Lower limits

Column 1: End $j = B$ is taken at the fixed end: $R_B = 1$.

End $j = A$ is taken at the beam location, where it is restrained by the beam ($b = 6$), but not by column 2 (pinned far end, $b = 0$). Total relative first-order rotational stiffness (equation [3.21]) becomes:

$$\kappa_{\theta A} = \frac{k_{\theta A}}{(EI_2/L_2)} = \frac{(6EI_b/L_b)}{(EI_1/L_1)} = \frac{6}{1} = 6,$$

which gives $R_A = 1/(1 + 2.4/6) = 0.714$.

Then, the lower limit (using equation [3.22]) becomes $\lim \bar{\beta}_{01} = 1.167$.

Column 2: End $j = B$ is taken at the pinned end: $R_B = 0$.

End $j = A$ is taken at the beam location, where it is restrained by the beam ($b = 6$) and column 1 ($b = 1$) with a total relative first-order rotational stiffness (equation [3.21]) of:

$$\kappa_{\theta A} = \frac{k_{\theta A}}{(EI_2/L_2)} = \frac{(6EI_b/L_b) + (1EI_1/L_1)}{(EI_2/L_2)} = \frac{6+1}{2} = 3.5,$$

and $R_A = 1/(1 + 2.4/3.5) = 0.593$.

Then, the lower limit (using equation [3.22]) becomes $\lim \bar{\beta}_{02} = 2.597$.

Step 1: "Conventional" effective lengths ($\kappa_{\theta j}$ by equation [3.27])

At the beam location ($j = A$), each column is restrained by a portion of the beam proportional to its EI/L value. This gives, for both columns, the same "conventional" relative restraint $\kappa_{\theta A}$ (equation [3.27]) and R_A:

$$\kappa_{\theta A} = \frac{\sum (bEI_b/L_b)}{\sum (EI/L)} = \frac{6EI_b/L_b}{(EI_1/L_1) + (EI_2/L_2)} = \frac{6}{1+2} = 2$$

which gives the fixity factor $R_A = 1/(1 + c/\kappa_{\theta A}) = 1/(1 + 2.4/2) = 0.455$.

Column 1: For $R_A = 0.455$ and $R_B = 1$ (at the fixed end), equation [3.22] gives $\beta_{01} = 1.375$.

Column 2: For $R_A = 0.455$ and $R_B = 0$ (at the pinned end), equation [3.22] gives $\beta_{02} = 2.967$.

Step 2 and 3b: Improved effective lengths

With $\beta_{01} = 1.375$ and $\beta_{02} = 2.967$, and the lower limits found above, the MOM yields for column $j = 1$ (equation [3.39]):

$$\left. \begin{aligned} &\bar{\beta}_{01} = \sqrt{\frac{\bar{\alpha}}{\alpha_{E1}}} = \left[\tfrac{1}{2}(\beta_{01}^2 + \beta_2^2 Q_{21}) \right]^{1/2} = 1.773 \\ &\bar{\beta}_{01} \geq \lim \bar{\beta}_{01} = 1.167 \\ &\bar{\beta}_{01} \geq (\lim \bar{\beta}_{02})\sqrt{Q_{21}} = 2.597\sqrt{0.5} = 1.836 \\ &\text{Thus, } \bar{\beta}_{01} = \underline{1.836} \text{ and} \\ &\bar{\beta}_{02} = \bar{\beta}_{01}\sqrt{Q_{12}} = 1.836\sqrt{2} = \underline{2.597} \end{aligned} \right\} \qquad [3.43]$$

The resulting $\bar{\beta}_0$ values are plotted in Figure 3.5(b), marked "Example", and can be seen to be in good agreement with the exact system instability values.

b) Application to braced quay-frame

To study the effect on the critical load of a braced frame to gradually increasing axial loads in one or more of its members, the frame in Figure 3.6(a) is considered. Exact results (from [FRI 94]) for the four-member frame defined in the figure are presented in Figure 3.7 in terms of β_{01} for member 1. β_0 factors for the other members can be obtained from equation [3.42] (remove bars above the factors).

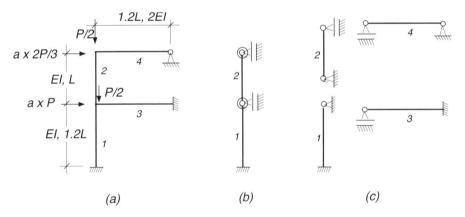

(a)   (b)   (c)

Figure 3.6. *Quay frame details*

The compression in the horizontal "beam" members (3 and 4) can be varied through the multiplier "*a*". This frame with $a = 1$ has been analyzed by the MOM before [HEL 96b], but then with less accurate lower limit constraints. The frame, considered to be an example of quay structure, was initially studied using another approach [BRI 86].

The axial forces in the members are taken as $N_1 = P$, $N_2 = P/2$, $N_3 = a \cdot P$ and $N_4 = a \cdot 2P$. As the member forces N_3 and N_4 increase with increasing multiplier, members 3 and 4 gradually "transform" from being restraining flexural members in the strict sense $(a = 0)$ to become "compression members".

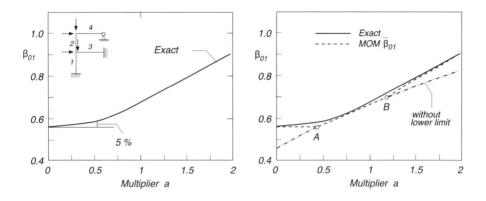

Figure 3.7. *Quay frame; (a) exact effective length and (b) exact versus predicted (MOM) effective length factors (from [HEL 97])*

With the increasing multiplier, β_{01} increases due to the reduced restraint afforded to member 1 by members 3 and 4. The increase is initially slow. At $a = 0$, members 3 and 4 are flexural beam members, with zero axial forces. The frame can then be represented by Figure 3.6(b), where the beams are represented by springs with first-order rotational stiffnesses. As seen in Figure 3.7(a), such a model can also be used for axial forces in these members up to a multiplier of about $a = 0.5$, provided an inaccuracy of about 5% is accepted.

However, for predictions by the MOM, all members will be classified as compression members $(n = 4)$ in order to demonstrate the robustness of the method (by use of "lower limits" on effective lengths).

Step 4: Lower limits

Member 1: The lower limit for member 1 can bedetermined from the subassemblage consisting of members 1, 2 and 3.

End $j = B$ is taken at the fixed base: $R_B = 1$.

End $j = A$ is taken at the top (junction with members 2 and 3).

First-order rotational restraint: member 2 is conservatively considered pinned at the far end ($b = 3$); member 3 is fixed at its far end ($b = 4$). Thus:

$$\kappa_{\theta A} = \frac{3(EI/L)_2 + 4(EI/L)_3}{(EI/L)_1} = 10, \quad R_A = \frac{1}{1 + 4.8/\kappa_{\theta A}} = 0.676$$

and, using equation [3.23] or [3.24], $\lim \bar{\beta}_{01} = 0.55$.

Member 2: It is restrained by the first-order rotational stiffness of members 1 and 3 at the lower end, and by member 4 at the top. This gives $\lim \bar{\beta}_{02} = 0.65$.

Member 3: It is fixed at the right end ($R_B = 1$) and restrained by the first-order rotational stiffness of members 1 and 2 at the left end. This gives $\lim \bar{\beta}_{0,3} = 0.58$.

Member 4: It is pinned at the right ($R_B = 0$) and restrained by first-order rotational stiffness of members 2 at the left (assumed pinned at the far end). This gives $\lim \bar{\beta}_{0,4} = 0.88$.

Step 1: "Conventional" effective lengths

When all members are assumed to be compression members, the conventional $\kappa_{\theta j}$ factors by equation [3.27] become zero at the interior joints, since there are no beam restraints "to be shared". The isolated elements in Figure 3.6(c) result.

Corresponding isolated, conventional β_0 factors become $\beta_{01} = 0.7$, $\beta_{02} = 1.0$, $\beta_{03} = 0.7$ and $\beta_{04} = 1.0$.

Step 2 and 3b: Improved effective lengths

With Q_{ij} defined by equation [3.40], the values $Q_{11} = 1$, $Q_{21} = 0.347$, $Q_{31} = 0.781a$ and $Q_{41} = 0.512a$, result.

Then, improved prediction of $\bar{\beta}_{01}$ for member $j = 1$ can be computed from equation [3.39]:

$$
\left.
\begin{aligned}
\bar{\beta}_{01} &= \left[\tfrac{1}{4}(\beta_{01}^2 + \beta_{02}^2 Q_{21} + \beta_{03}^2 Q_{31} + \beta_{04}^2 Q_{41})\right]^{1/2} \\
\bar{\beta}_{01} &\geq \lim \bar{\beta}_{01} = 0.55 \\
\bar{\beta}_{01} &\geq (\lim \bar{\beta}_{02})\sqrt{Q_{21}} = 0.65\sqrt{0.347} \\
\bar{\beta}_{01} &\geq (\lim \bar{\beta}_{03})\sqrt{Q_{31}} = 0.58\sqrt{0.781a} \\
\bar{\beta}_{01} &\geq (\lim \bar{\beta}_{04})\sqrt{Q_{41}} = 0.88\sqrt{0.521a}
\end{aligned}
\right\}
\qquad [3.44]
$$

Once $\bar{\beta}_{01}$ is determined, those for the other members can be determined from equation [3.42]:

$$
\bar{\beta}_{0i} = \bar{\beta}_{01}\sqrt{Q_{1i}} = \frac{\bar{\beta}_{0j}}{\sqrt{Q_{i1}}} \qquad i = 2, 3, 4
$$

The MOM predictions for $\bar{\beta}_{01}$ are shown in Figure 3.7. Predictions are in excellent agreement with exact results for all a values considered. The lower limit on member 1 ($\lim \beta_{01} = 0.55$) governs to the left of point A and the lower limit on member 4 (last equation above) to the right of point B. The lower limits for members 2 and 3 do not govern for any a values.

For a specific case, it might be clear from the start that some members, with negligible axial forces, may be considered restraining (beam), and that members are clear compression members. In such cases, it may not be necessary to compute lower limits at all.

3.3. Global buckling of unbraced or partially braced systems

3.3.1. General considerations

3.3.1.1. Column interaction and pseudo-critical loads

In unbraced moment frames, in which lateral stiffness is primarily provided by the columns, the stiffer ("stronger") columns will interact with the more flexible ("weaker") columns, and will, through this interaction, provide lateral support to the more flexible columns in order to maintain overall system stability. Similarly, also for frames with partial lateral restraint. The effective length formulas for individual columns considered previously cannot be used directly to provide effective lengths for columns in such frames, but they can be used "indirectly", as shall be seen.

In this section, most attention is given to single-level (story) frames. The frame can also be a story isolated from a multi-story structure, using, for instance, the conventional vertical interaction assumption. Such an application is considered in section 3.3.6.

For later use, it is convenient to include subscripts "s" and "b" to denote effective lengths for the free-sway and fully braced case of individual columns (Figure 3.2), respectively. Thus, $L_{0s} = \beta_{0s} L$ and $L_{0b} = \beta_{0b} L$. The corresponding individual critical loads become:

$$N_{cs} = \frac{\pi^2 EI}{(\beta_{0s} L)^2} \quad \text{and} \quad N_{cb} = \frac{\pi^2 EI}{(\beta_{0b} L)^2} \qquad [3.45]$$

and the corresponding load indices:

$$\alpha_s = \frac{N}{N_{cs}} = \frac{N}{N_E} \beta_{0s}^2 \quad \text{and} \quad \alpha_b = \frac{N}{N_{cb}} = \frac{N}{N_E} \beta_{0b}^2 \qquad [3.46]$$

These individual unbraced critical loads are strictly pseudo-critical loads, which may deviate to various extents from real critical loads due to errors in restraint conditions. For instance, in sway frames, individual members are not free to deflect freely, but are forced to act together during side-way.

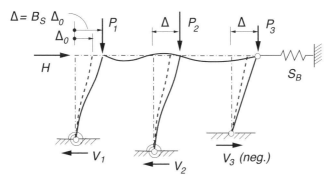

Figure 3.8. *Partially braced multi-bay frame (from [HEL 09a])*

3.3.1.2. *Story sway magnification and system instability*

Similar to that of a single column (Chapter 2, equation [2.15]), a story sway magnifier B_s for a frame with side-way, whether unbraced or partially braced, such as the one in Figure 3.8, is defined as the ratio between the total and the first-order lateral displacement. It can be given by:

$$B_s = \frac{\Delta}{\Delta_0} = \frac{1}{1 - \alpha_{ss}} \qquad [3.47]$$

where α_{ss} is the system (story) sway stability index.

B_s will approach infinity, as the loading approaches the critical loading and α_{ss} approaches 1.0. Consequently, the system (story) instability condition is defined $\alpha_{ss} = 1$.

3.3.1.3. *System (story) sway stability index*

Assuming that the axial deformations in the connecting beam (girder) is negligible, the system (story) sway stability index can be expressed by [HEL 09a]:

$$\alpha_{ss} = \frac{\sum (\gamma_n N/L)}{(H/\Delta_0)} \qquad [3.48]$$

or by

$$\alpha_{ss} = \frac{\sum (\gamma_n N/L)}{\sum (\gamma_s N_{cs}/L) + S_B} \qquad [3.49]$$

where the summations are over all interconnected columns, H and Δ_0 are the applied lateral load at the story top and the corresponding relative first-order lateral displacement (between bottom and top), respectively, $N_{cs} (= N_{cr,unbraced})$ is a free-sway critical load as defined above and γ_s and γ_n are flexibility factors at free-sway and, in general, reflecting the increased lateral flexibility of an axially loaded member as compared to one without axial load. The basis and approximations of the γ factors will be defined and discussed in the following.

3.3.1.4. *Comments on derivations*

The most general derivations are based on equilibrium considerations [HEL 76, HEL 09a]. A simpler derivation for the first α_{ss} expression can be obtained by introducing a fictitious horizontal load $H_{eq} = \sum (\gamma_n N \Delta/L)$ that has the same effect on the frame side-way as the axial loads. The γ factor corrects for the fact that a member with a nonlinear moment distribution, caused by an axial load, is more flexible than a member with a linear moment distribution, caused by a horizontal load. The lateral displacement is proportional to the applied lateral load. This gives the relationship:

$$\frac{\Delta}{\Delta_0} = \frac{H + H_{eq}}{H} = \frac{H + \sum(\gamma_n N \Delta/L)}{H} \qquad [3.50]$$

By solving for Δ/Δ_0, B_s with α_{ss} defined by equation [3.48] is obtained.

The applied lateral load H will be resisted by the sum of shear forces in the columns plus the bracing force ($S_B \Delta_0$). It should be recognized that H/Δ_0 is the first-order lateral story stiffness (force per unit displacement). Provided the first-order analysis include shear strains, H/Δ_0 will include effects of shear strain deformations in the columns, and, if the bracing force is due to a shear wall, in the shear wall as well. Normally, shear strain effects in columns are neglected.

In the second expression, equation [3.49], shear strain deformations in columns are not accounted for. However, if a shear wall contributes to the bracing, S_B may include average shear strain effects through $S_B = GA/L$, where G is the shear modulus and A is the shear area in direction of the displacement. The flexibility factors for a shear wall may be taken as $\gamma_n = \gamma_s = 1$.

Provided shear deformations are neglected, both of the α_{ss} expressions above will give exactly the same results provided they both are based on the same rotational end restraints at the base and top of each column in the summations. Equation [3.48] implies first-order restraints. First-order restraint assumptions will generally be adequate [HEL 09b]. Exceptions exist, however, such as in single curvature regions of multi-story frames, where columns on one level may contribute significantly to the restraint of columns on adjacent levels. Axial load effects on restraints may be significant in such cases. Different restraint assumptions, and possible simplifications, will introduce differences in the two formulations. Which formulation to use, is often a question of type of application, and sometimes preference.

Equation [3.49], first derived for the case with $\gamma_n = \gamma_s$ and $S_B = 0$ [HEL 76], is often a convenient form, in particular when first-order analysis results are not available. The formulations above, with a general γ_n factor, introduced quite recently [HEL 09a], allowed for the first time the reflection of the full sway-braced column interaction (including columns that are bracing to columns that are partially, or nearly fully, braced by the others) in story magnifier formulations.

3.3.1.5. *Single curvature regions*

The presented sway magnifiers are primarily suited for application to stories in which the columns bend in double curvature, with an inflection point (zero moment) either between or at a column end. For such frames, use of first-order end restraints of columns will generally be acceptable. This is implied when taking α_{ss} according to equation [3.48] and may be assumed when using equation [3.49] (in the effective length computations). For stories located in single curvature regions of multi-story frames, second-order load effects may significantly affect end restraints. An approach whereby this may be accounted for in a simple manner, by introducing a "combined" flexibility factor γ_c, has been proposed [HEL 09b], and will be briefly presented in section 3.4.4.

3.3.2. *Flexibility factors*

3.3.2.1. *Flexibility factor approximations*

Possible γ approximations, to various degrees of accuracy, can be defined by:

$$
\begin{array}{lll}
(a) & \gamma_n = \gamma_s = 1 & \\
(b) & \gamma_n = \gamma_s = 1.1 - 1.15 & \\
(c) & \gamma_n = \gamma_s \quad (\text{equation } [3.52]) & [3.51] \\
(d) & \gamma_n = \gamma_s + \gamma_1 \quad (\text{equation } [3.53a] \text{ with } \gamma_2 = 0) & \\
(e) & \gamma_n = \gamma_s + \gamma_1 + \gamma_2 \quad (\text{equation } [3.53a]) &
\end{array}
$$

One approximation may be found acceptable in one case, and not in another.

3.3.2.2. *Flexibility factor* γ_s

The γ_s factor reflects the increased lateral flexibility of an axially loaded unbraced member (with a nonlinear moment distribution along the member) as compared to that of a member with a (first-order) linear moment distribution. At the free-sway critical load, which is the state of interest here, it varies between 1 to 1.216 (1.22) for positive rotational end restraints. Its variation is shown in Figure 3.9 for various combinations of positive and negative rotational end restraints. They are expressed in terms of the smaller (R_{MIN}) and the larger (R_{MAX}) of the *degree of rotational fixity factors* R_A and R_B at member ends defined in the figure (in which $k_j = k_{\theta j}$) or by equation [3.19] with $c = 2$.

An approximate, yet rather accurate, γ_s factor can be given in alternative forms by:

$$\gamma_s = 1 + \frac{L^4}{167(EI\Delta_0)^2} \left[M_{0A}M_{0B} + (M_{0A} - M_{0B})^2 \right] \qquad [3.52a]$$

$$\gamma_s = 1 + 0.216 \frac{R_A R_B + 4(R_A - R_B)^2}{(R_A + R_B - 3)^2} \qquad [3.52b]$$

$$\gamma_s = 1 + 0.216 \frac{(G_A + 3)(G_B + 3) + 4(G_A - G_B)^2}{[(G_A + 2)(G_B + 2) - 1]^2} \qquad [3.52c]$$

The first factor is written in terms of first-order moments and displacements, and is similar too, but a modified version of an expression is given by Rubin [RUB 73]. The second factor is written in terms of first-order fixity factors (equation [3.19] with $c=2$), and the third in terms of G factors (equation [3.14] with $b_o = 6$). The basis for these factors and comparisons with exact results are given in [HEL 09a, HEL 09b]. The accuracy of equation [3.52] is very good (normally within a fraction of a percent) for various combinations of positive and negative end restraints.

Alternative approximate factors, reviewed in [HEL 08a], have been given in [HEL 76, LEM 77, LUI 92] or implied in other expressions [ARI 95].

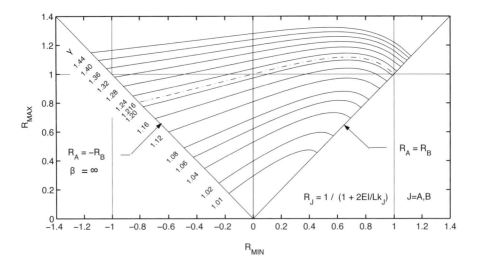

Figure 3.9. *Variation of the flexibility factor $\gamma = \gamma_s$ (at the free-sway condition) in terms of fixity factors of the end restraints (from [HEL 09a])*

3.3.2.3. *Extended flexibility factor γ_n*

The flexibility factor γ_n is load dependent and accounts for the increased flexibility, beyond that reflected by linear moment distributions, of both unbraced and partially to fully braced axially loaded members. It varies between 1 and large values, as the member varies from unbraced member towards an increasingly braced column. An approximation proposed in [HEL 09a], based on a study of the mechanics of the response for a variety of parameters, is given by:

$$\gamma_n = \gamma_s + \gamma_1 + \gamma_2 \quad (\geq \gamma_s) \tag{3.53a}$$

$$\gamma_1 = 0.12\,(\gamma_s - 1)(\alpha_s - 1) \tag{3.53b}$$

$$\gamma_2 = 0.6\,\alpha_{s,b}\left(\frac{\alpha_s - 1}{\alpha_{s,b}}\right)^8 \tag{3.53c}$$

where

$$\alpha_s = \frac{N}{N_{cs}} = (\alpha_E \beta_{0s}^2) \quad \text{and} \quad \alpha_{s,b} = \frac{N_{cb}}{N_{cs}} = \left(\frac{\beta_{0s}}{\beta_{0b}}\right)^2 \qquad [3.54]$$

Here, α_s is the unbraced (free-to-sway) load index (as defined before) and $\alpha_{s,b}$ is the load index defined with an axial load equal to the pseudo-critical braced load, that is $N = N_{cb}$.

The lower limit (γ_s) given in parenthesis in equation [3.53a] may be adopted when this represents a simplification. For pinned-end columns ($N_{cs} = 0$), $\gamma_n = 1$. The same value is conservatively recommended for the occasional, rare column with negative (tensile) axial loads and for possible shear walls.

Case a in equation [3.51] is generally too approximate and is not recommended except at very low axial load levels. For laterally stiffening columns ($\alpha_s < 1$), case c will give representative predictions. Case d is applicable to stiffened columns ($\alpha_s > 1$) with low to moderately high load levels for some end restraint combinations, and for very high load levels for columns with nearly equal end restraints. Case e should be used for columns with high axial load levels, approaching the columns local, braced buckling load.

Simplifications are justified in a large number of practical cases. For practical unbraced frames with reasonably similar columns case b with $\gamma_n =$ constant in the range 1.1–1.15, will normally be acceptable. Additional discussions and details are given in [HEL 09a].

3.3.3. *System instability and "system" effective lengths*

Rigorous stability analysis will detect both buckling of individual members in local modes and buckling of the system in global, or overall, modes that include all members. In approximate methods, it is necessary to consider local and global (system) modes separately. The presentation below follows in the main aspects the presentation given in [HEL 09a].

3.3.3.1. *Local column instability*

Local instability of an individual column will fail in a mode similar to that of a fully braced column. An approximate indication of local buckling, then, may be obtained considering the braced member stability index, which, for a given column k, is defined by:

$$\alpha_{b,k} = \frac{N_k}{N_{cb.k}} = \frac{N_k}{N_{Ek}} \beta_{0b,k}^2 \qquad [3.55]$$

Here, N_k is the axial force in column k due to a given ("reference") loading, $N_{cb,k} (= N_{cr\,braced})$ is the braced buckling load of column k, and $\beta_{0b,k}$ is the corresponding effective length factor (equation [3.45]). Let us assume that, of all the interconnected columns "n" on the same level (story), column k has the greatest braced stability index:

$$\alpha_{b,k} = \max(\alpha_{b,i}) \quad i = 1, 2, 3..., n \qquad [3.56]$$

and therefore is the most vulnerable to local buckling.

To reflect the fact that a column in a sway frame will not be completely braced, a somewhat larger value may be adopted in practical calculations. A value of $a^2 \alpha_{b,k}$, with a in the range 1.05–1.1 has been suggested [HEL 09a]. Alternatively, the MOM could have been used to obtain improved estimates.

3.3.3.2. *System (story) instability*

It will be assumed that the axial load in the various columns is little affected by second-order load effects, such that they can be taken equal to those (N) obtained from a first-order analysis for a given ("reference") loading. Increasing column axial loads can then be expressed through a single load factor λ_f, which is applied to the axial ("reference") loads N_i in each column axis i.

The system (story) stability condition is obtained, when, for increasing loads in the various columns axes, the system stability index α_{ss} becomes equal to unity. For simplicity, only the most common case of proportional loading is considered. For clarity, cases with load-independent and load-dependent flexibility factors γ are considered separately.

a) *Simplified approach when $\gamma_n = \gamma_s$ (or some constant)*

When the flexibility factor $\gamma_n = \gamma_s$, or some other constant, independent of the axial column load, the load factor required to give instability of the total system (story), $\lambda_{f\,cr}$, is simply equal to the inverse of the α_{ss} value as calculated with the reference loading. Thus, when also considering possible local member buckling, the governing load index becomes:

$$\alpha = \max \left(\alpha_{ss},\ a^2\, \alpha_{b,k} \right) \tag{3.57}$$

The critical axial load in a column i then becomes:

$$N_{cr,i} = \lambda_{f\,cr} N_i = \frac{N_i}{\alpha_{ss}} \leq \lim N_{cr,i} \tag{3.58}$$

where

$$\lim N_{cr,i} = \frac{N_i}{a^2\, \alpha_{b,k}} \tag{3.59}$$

The corresponding "story" effective length factor for column "i" can be expressed by:

$$\beta_{0i} = \left[\frac{N_{E\,i}}{N_{cr,i}} \right]^{\frac{1}{2}} \tag{3.60}$$

or, when substituting for $N_{cr,i}$, by

$$\beta_{0i} = \left[\frac{N_{E\,i}}{N_i} \alpha_{ss} \right]^{\frac{1}{2}} \geq a \cdot \lim \beta_{0i} \tag{3.61}$$

where

$$\lim \beta_{0i} = \left[\frac{N_{E\,i}}{N_i} \alpha_{b,k} \right]^{\frac{1}{2}} = \beta_{0b,k} \left[\frac{\alpha_{E\,k}}{\alpha_{E\,i}} \right]^{\frac{1}{2}} \tag{3.62}$$

The critical loads, and the effective lengths, of the columns of the story are interrelated through the common α_{ss} or $\alpha_{b,k}$ factor. Prediction accuracies will consequently be the same for all the columns, also when the local buckling limits apply. Strengthening of the weakest column(s) is the most efficient way to increase the loading causing system instability.

b) *Extended approach with* $\gamma_n = \gamma_n(N)$

When the flexibility factor is taken according to the extended γ_n formulation (equation [3.53a]), the α_{ss} expressions (equations [3.48] and [3.49]) also become nonlinear functions of the axial load. In this case, an explicit solution of the critical system instability loading $N_{cr,j}$, giving $\alpha_{ss} = 1$, can not be obtained.

The most practical way for this case is to use an iterative solution strategy:

– Let N_i be the initial (reference) load in the respective columns, and replace N_i in the α_{ss} expressions by $\lambda_f N_i$, where λ_f is the common load factor (proportional loading).

– Compute α_{ss} for increasing loads $\lambda_f N_i$ until $(1 - \alpha_{ss})$ becomes zero (vanishing of the stiffness matrix determinant), as illustrated schematically in Figure 3.10.

– The loading at this stage is the critical loading: $N_{cr,i} = \lambda_{fcr} N_i$.

The $(1 - \alpha_{ss})$ *versus* λ_f variation (Figure 3.10) is rather linear in practical cases. Therefore, only a few load steps, combined with extrapolation, or interpolation, as the case may be, are normally sufficient for determining the critical load factor with satisfactory accuracy. As this factor is approached, the sway magnifier B_s approaches infinity.

The critical load for a column i, when also considering possible local buckling, is then given by:

$$N_{cr,i} = \lambda_{fcr} N_i \leq \lim N_{cr,i} \tag{3.63}$$

where $\lim N_{cr,i}$ is given by equation [3.59]. The corresponding effective length factor becomes:

$$\beta_{0i} = \left[\frac{N_{Ej}}{N_{cr,i}}\right]^{\frac{1}{2}} \tag{3.64}$$

or, alternatively,

$$\beta_{0i} = \left[\frac{N_{Ei}}{\lambda_{fcr}N_i}\right]^{\frac{1}{2}} \geq a \cdot \lim \beta_{0i} \tag{3.65}$$

where $\lim \beta_{0i}$ is given by equation [3.62].

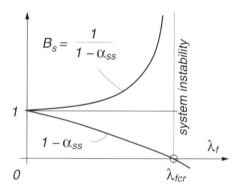

Figure 3.10. *Sway magnifier and system instability loading*

c) *"Semi-extended" approach with* $\gamma_n = \gamma_1 + \gamma_2$

In this case, $\gamma_n = \gamma_1 + \gamma_2$ becomes a linear equation in the axial load (rather than a nonlinear function in the general case). It is then strictly not necessary to iterate, as a quadratic equation in λ_{fcr} can be obtained, from which λ_{fcr} can be solved explicitly. An example of such a case is considered in section 3.3.4 (case 2).

3.3.3.3. *Comments on practical use*

A system instability analysis gives important information of the critical loading for the total system. This is frequently used to establish overall sway and moment magnifiers for frames. As far as individual column information from a system analysis is concerned, such as column critical load and corresponding effective length, it is of most use for the column, or a cluster of strongly interacting columns, that contributes most to system instability ("weakest link(s)"). By increasing the stiffness of these ("weakest links"), the system critical load can be increased substantially in many instances.

The systems critical load can be expressed, as seen, in terms of the critical load (or effective length) of any column in the system. Although the stiffer columns also reach, per definition, their "critical loads" at the same instant as the whole system becomes unstable, a change of their stiffness (up or down) may have little effect on the final result. For such stiffer, "stronger", columns, the system critical load information may not be too useful. Exception are such columns when they interact strongly with more flexible ("weaker") columns, and thus delay frame instability through the stiffening effects of such interaction. Lateral column interaction between columns on the same level of a sway frame is an example of such a case (see section 3.3.5).

3.3.4. *Instability of partially braced column – example*

The laterally restrained column shown by the inset in Figure 3.11 is considered. It is fixed at the base and pinned at the top. It has been considered before [HEL 09a], and represents a severe test case due to the large difference in rotational restraints at the two ends. Exact solutions (full lines) are obtained using convenient expressions given in [CHE 99]. Critical loads are zero when the lateral bracing stiffness is equal and opposite to the columns own first-order lateral stiffness. The column can be considered to be a part of an unbraced or partially braced frame, in which the stiffness S_B of the lateral spring bracing represents the interaction with other columns and partial bracing.

The system sway stability index α_{ss} will first be taken according to equation [3.48]. The first-order lateral stiffness of the cantilever column alone is $3EI/L^3$. Thus, the denominator becomes:

$$H/\Delta_0 = (3EI/L^3) + S_B$$

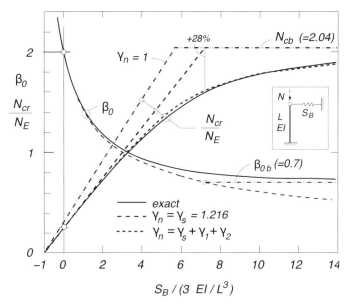

Figure 3.11. *Comparison of approximate and exact critical loads and effective
lengths for partially braced fixed-pinned column*

If instead, α_{ss} by equation [3.49] had been used, with $\gamma_s = 1.216$ (equation
[3.52b] for $R_B = 1$ and $R_A = 0$) and $N_{cs} = \pi^2 EI/(2L)^2$, the exact same
denominator is obtained. Then:

$$\alpha_{ss} = \frac{(\gamma_n N/L)}{(H/\Delta_0)} = \frac{(\gamma_n N/L)}{3\frac{EI}{L^3} + S_B} \tag{3.66}$$

Column parameters: $\beta_{0s} = 2$, $\beta_{0b} = 0.7$, $\alpha_{s,b} = (2/0.7)^2 = 8.163$, $\alpha_s = N/N_{cs} = \alpha_E \beta_{0s}^2 = 4\alpha_E$.

Four γ_n cases will be considered:

1) $\gamma_n = 1$

2) $\gamma_n = \gamma_s = 1.216$

3) $\gamma_n = \gamma_s + \gamma_1 = 1.216 + 0.026\,(4\alpha_E - 1)$

4) $\gamma_n = \gamma_s + \gamma_1 + \gamma_2 = 1.216 + 0.026\,(4\alpha_E - 1) + 4.898\,[(4\alpha_E - 1)/8.16]^8$

For cases 1 and 2, with constant γ_n, the critical load given by equation [3.63] becomes:

$$N_{cr} = \frac{N}{\alpha_{ss}} = \frac{1}{\gamma_n} \cdot \frac{3EI}{L^2} \left(1 + \frac{S_B}{3EI/L^3}\right) \le \frac{N_{cb}}{a^2}$$

or, in non-dimensional form:

$$\alpha_{E\,cr} = \frac{N_{cr}}{N_E} = \frac{1}{\gamma_n} \cdot \frac{3}{\pi^2} \left(1 + \frac{S_B}{3EI/L^3}\right) \le \frac{1}{a^2 \beta_{0b}^2} \qquad [3.67]$$

The critical load is linearly proportional to the lateral restraint S_B. Results for cases 1 and 2 are plotted in Figure 3.11. This is an extended version of a similar figure presented before [HEL 09a]. The local buckling limit (horizontal line) is plotted for the uncertainty factor $a = 1$.

For the simplified approach in the simplest case with $\gamma_n = 1.0$, the accuracy cannot be considered acceptable except for very small critical axial load levels (obtained with low bracing values).

For $\gamma_n = \gamma_s = 1.216$, the accuracy is acceptable at low and intermediate lateral bracing values, but not for high values at which the buckling mode approaches braced buckling. The corresponding effective length factor, given by equation [3.64], is also shown in the figure.

In case 4, with the extended γ_n, N_{cr} can be determined iteratively as described above. However, since the system involves a single column only, it is easier simply to specify $N = N_{cr}$ and compute the corresponding bracing stiffness S_B from equation [3.66] for $\alpha_{ss} = 1$. This gives:

$$\bar{S}_B = \frac{S_B}{3EI/L^3} = \frac{\pi^2}{3} \cdot \gamma_n \frac{N_{cr}}{N_E} - 1 \qquad [3.68]$$

Sample calc. 1. $\alpha_{E\,cr} = N_{cr}/N_E = 0.75$, $\gamma_n = 1.216 + 0.052 + 0.000 = 1.268$:

$\bar{S}_B = 2.13$

Sample calc. 2. $\alpha_{E\,cr} = N_{cr}/N_E = 1.2$, $\gamma_n = 1.216+0.099+0.011 = 1.326$: $\bar{S}_B = 4.23$ ($\bar{S}_B = 4.19$)

Sample calc. 3. $\alpha_{E\,cr} = N_{cr}/N_E = 1.5$, $\gamma_n = 1.216+0.130+0.097 = 1.443$: $\bar{S}_B = 6.12$ ($\bar{S}_B = 5.64$)

Sample calc. 4. $\alpha_{E\,cr} = N_{cr}/N_E = 1.8$, $\gamma_n = 1.216+0.162+0.544 = 1.922$: $\bar{S}_B = 10.38$ ($\bar{S}_B = 7.57$)

The numbers in parentheses are the $\bar{S}_B$ values obtained for γ_n-case 3, that is they are the values obtained without the last term (γ_2) in the γ_n calculations above. For $\alpha_{E\,cr} = N_{cr}/N_E < 1.5$, or so, the accuracy is still quite acceptable. It is clear from these results that γ_2 can be neglected up to fairly high load levels without losing too much accuracy.

The agreement with the exact critical results is generally seen to be very good when the extended γ_n is used. The accuracy has been found [HEL 09a] to be generally well within $\pm 2\%$ ($\pm 1\%$ in corresponding effective lengths) for a range of combinations of rotational restraint at the column base and top.

3.3.5. Instability of partially braced frame – example

3.3.5.1. Frame and loading

The three-bay bridge (or single-story frame) in Figure 3.12 is considered. Column 4 of the knee-frame at the right end of the frame has negligible axial load and provides lateral bracing of the three other columns.

Assumed section stiffnesses and lengths:

$EI_1 = EI$, $EI_2 = EI$, $EI_3 = 0.5EI$, $EI_4 = EI$, $EI_b = EI$.

$L_1 = L$, $L_2 = L$, $L_3 = 1.5L$, $L_4 = L$.

First-order axial forces N_i:

$N_1 = P_1 = 0.5P$, $N_2 = P_2 = P$, $N_3 = P_3 = 2P$, $N_4 = P_4 = 0$.

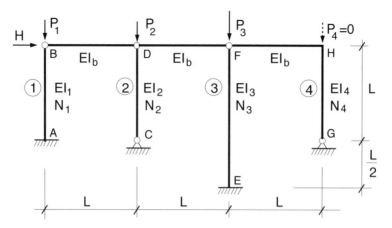

Figure 3.12. *Sway frame example*

Column 3, with the larger axial force and length, and the smaller section stiffness, is clearly the most flexible of the columns. The shorter column 1 with a small axial load will be considerably stiffer. Column 2, pinned at both ends, is a column that leans on the others for lateral stability.

3.3.5.2. *Local (braced) instability*

Braced effective length factors: $\beta_{0b,1} = 0.7$, $\beta_{0b,2} = 1$, $\beta_{0b,3} = 0.7$.

Braced pseudo-critical loads, $N_{cb,i} = \pi^2 EI_i / (\beta_{0b,i} L_i)^2$: $N_{cb,1} = 2.04 N_E$, $N_{cb,2} = N_E$, $N_{cb,3} = 0.454 N_E$. For simplicity, these critical loads have been expressed in terms of:

$$N_E = \pi^2 EI/L^2$$

Braced load indices: $\alpha_{b,1} = N_1/N_{cb,1} = 0.245 P/N_E$; $\alpha_{b,2} = N_2/N_{cb,2} = 1.0 P/N_E$; $\alpha_{b,3} = N_3/N_{cb,3} = 4.41 P/N_E$. Clearly, column 3 is the column that is most vulnerable to local buckling.

3.3.5.3. *System (story) instability*

First-order lateral stiffness H/Δ_0: column 2 (pinned-end) does not contribute. Columns 1 and 3 have $3EI_i/L_i^3$, and for column 4, restrained at the top by a beam, $3EI_4/2L_4^3$ can be found. Totally, in terms of EI and L:

$$H/\Delta_0 = 89EI/18L^3 = 0.501 N_E/L \qquad [3.69]$$

Numerator of the system stability index:

$$\sum \frac{\gamma_n N}{L} = \frac{\gamma_{n1} N_1}{L_1} + \frac{\gamma_{n2} N_2}{L_2} + \frac{\gamma_{n3} N_3}{L_3} = \left(0.5\gamma_{n1} + \gamma_{n2} + \frac{2\gamma_{n3}}{1.5}\right)\frac{P}{L}$$

[3.70]

Case 1: $\gamma_n = \gamma_s$

Flexibility factors (equation [3.52b]):

Columns 1 and 3, with fixed base ($R_B = 1$) and pinned top ($R_A = 0$):

$\gamma_{s1} = 1.216$ and $\gamma_{s3} = 1.216$.

Column 2, pinned at both ends: $\gamma_{s2} = 1$.

Then, equation [3.70] and the system stability index become:

$$\sum \frac{\gamma_n N}{L} = 3.229 \frac{P}{L} \quad \text{and} \quad \alpha_{ss} = \frac{\sum(\gamma_n N/L)}{H/\Delta_0}$$

$$= \frac{3.229 P/L}{0.501 N_E/L} = 6.445 \frac{P}{N_E}$$

Check local buckling:
$\alpha_{ss} > \max(\alpha_{b,i}) = \alpha_{b,3} = 4.41 P/N_E$. Thus, local buckling does not govern.

Critical (system) loads in the columns: $N_{cr,i} = N_i/\alpha_{ss}$:
$N_{cr,1} = 0.5P/(6.445P/N_E) = 0.0776 N_E$;
$N_{cr,2} = P/(6.445P/N_E) = 0.155 N_E$;
$N_{cr,3} = 2P/(6.445P/N_E) = 0.310 N_E$.

Effective length factors $\beta_{0i} = \sqrt{N_{Ei}/N_{cr,i}}$:

Noting that $N_{E1} = N_E$, $N_{E2} = N_E$ and $N_{E3} = (0.5/1.5^2)N_E = 0.222N_E$,

$$\beta_{01} = \sqrt{\frac{1}{0.0776}} = 3.59; \quad \beta_{02} = \sqrt{\frac{1}{0.155}} = 2.54;$$

$$\beta_{03} = \sqrt{\frac{0.222}{0.310}} = 0.846$$

Case 2: $\gamma_n = \gamma_s + \gamma_1$ *(load dependent)*

As before, $\gamma_n = 1$ for column 2. For columns 1 and 3, $\gamma_s = 1.216$ and $\gamma_{ni} = \gamma_{si} + 0.12\,(\gamma_{si} - 1)(\alpha_{si} - 1) = 1.190 + 0.104N_i/N_{Ei}$.

Substitution into equation [3.70] and collection of terms give:

$$\sum \frac{\gamma_n N}{L} = (3.182 + 1.301\frac{P}{N_E})\frac{P}{L} \quad \text{and}$$

$$\alpha_{ss} = \frac{\sum \gamma_n N/L}{H/\Delta_0} = (6.351 + 2.597\frac{P}{N_E})\frac{P}{N_E}$$

By setting $\alpha_{ss} = 1$ (instability condition), a quadratic equation in P_{cr}/N_E results. The quadratic equation can be solved explicitly, which is an advantage of the $\gamma_n = \gamma_s + \gamma_1$ formulation (linear in the axial loads) as compared to $\gamma_n = \gamma_s + \gamma_1 + \gamma_2$ (nonlinear in the axial loads).

By solving the quadratic equation, $P_{cr}/N_E = 0.1484$ is obtained.

Critical (system) loads in the columns:

$N_{cr,1} = 0.5P_{cr} = 0.0742N_E$; $N_{cr,2} = P_{cr} = 0.148N_E$;

$N_{cr,3} = 2P_{cr} = 0.297N_E$.

Effective length factors $\beta_{0i} = \sqrt{N_{Ei}/N_{cr,i}}$:

$$\beta_{01} = \sqrt{\frac{1}{0.0742}} = 3.67\,; \quad \beta_{02} = \sqrt{\frac{1}{0.148}} = 2.60\,;$$

$$\beta_{03} = \sqrt{\frac{0.222}{0.297}} = 0.865$$

Accuracy: Exact results, computed using stiffness relationships with stability functions, are found to be $N_{cr,3} = 0.290\,N_E$ and $\beta_{03} = 0.875$ for column 3. The approximate critical load and effective length factor results in case 2 above, are 2.4% greater and 1.1% smaller, respectively, than the exact results. This is only slightly to the unsafe-side, and is considered to be very good. Use of the more accurate $\gamma_n = \gamma_s + \gamma_1 + \gamma_2$ relationship would have given almost full agreement with exact results.

The comparable errors in the approximate results in case 1 ($N_{cr,3} = 0.310N_E$ and $\beta_{03} = 0.846$) obtained with $\gamma_n = \gamma_s$, are +6.9% and -3.3%, respectively. Although not quite as good as in case 2, this is also a very acceptable accuracy.

3.3.5.4. *Comments*

The effective length factor (greater than 1.0) of the pinned-end column 2 may look strange. However, when interpreting the effective length factors from system instability analyses, it should be recalled that system instability can be expressed through the critical load, and effective length, of any column. So also in principle for the pinned-end, "leaning" column 2, for which the real physical effective length factor is equal to the "local" value of $\beta_{02} = 1$ (corresponding to the distance between pinned ends (points with zero moments)).

To reduce confusion, it might be advisable not to give "system values" for columns that can be considered in isolation. For such columns, the "system" value ($\beta_{02} = 2.54$ in this case) gives no useful *member* information. The effective length to be used in design of the column (for imperfections, for instance) is its real local value $\beta_{02} = 1$. The isolated, real critical load of the column is $N_{cr,2} = N_E$, which is 6.45 times the value obtained from the system analysis. Without affecting the system critical load, the section stiffness of the column could in other words be reduced from EI to $EI/6.45$.

The large effective length factor of column 1 ($\beta_{01} = 3.67(3.59) > \beta_{0s,1} = 2$) reflects that it is a stiffening ("leaned-to") column. It provides significant lateral restraint to the frame. As an effect of this, the resultant force at column top (from the vertical, applied force and the horizontal force transmitted through the beam) will be inclined toward the left as schematically indicated in Figure 3.13. The thrust line will intersect the deflection shape at the top, and at the mathematical continuation below the column base. At the intersection points, moments are zero and the distance between these points are per definition the effective length.

For column 3, with an effective length factor ($\beta_{03} = 0.85$) that is considerably smaller than the free-sway factor ($\beta_{0s,3} = 2$), the opposite is the case. The column receives considerable lateral restraint from columns 1 and 4. This restraint force and the applied vertical load combine to a resulting force that is inclined toward the right. The resulting thrust line intersects the deflection shape within the column length as indicated in the figure. Column 3 is a stiffened ("leaning") column.

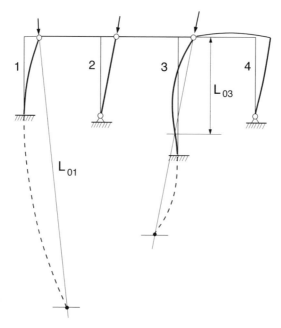

Figure 3.13. *Sway frame example – resulting effective lengths*

3.3.6. *Sway buckling of unbraced multistory frames*

3.3.6.1. *Method of means for multistory frames*

Buckling of larger frame systems is often initiated in limited regions that may be significantly more flexible than the rest of the frame. In section 3.2, it was suggested, therefore, that the MOM should be applied to a limited number of interacting columns in a limited region of the frame, in order for the method not to "suppress" localized failure in a region. For braced frames, a partial application to a limited number of *interconnected* compression members is probably most practical and appropriate. For multi-story sway frames, this may similarly apply to a limited number of stories.

The lateral (horizontal) interaction between columns of various stiffness in a story of an unbraced frame, resulting from the common translational displacement (Δ) of the top relative to the base of the columns, has been dealt with above. This approach may be applied to each of the stories in an unbraced frame as described in [HEL 98].

The story stability index defined in terms of pseudo-critical column loads is adopted. For a typical story i in a regular building with a constant story height L, it (equation [3.49]) can be simplified to:

$$\alpha_{ss,i} = \frac{\sum \gamma_n N}{\sum \gamma_s N_{cs}} \approx \frac{\sum \gamma_s N}{\sum \gamma_s N_{cs}} \approx \frac{\sum N}{\sum N_{cs}} \qquad [3.71]$$

To calculate this for individual stories, it is necessary to "divide" the stiffnesses of the restraining beams meeting at a joint, to the column below and above the joint. The conventional end restraint stiffness definition (equation [3.33]) may be adopted for this purpose. It results in the previously discussed restraint at a column end defined by equation [3.27].

$$\kappa_{\theta j} = \frac{k_{\theta j}}{(EI/L)} = \frac{1}{f_{\theta j}} = \frac{\sum (b\, EI_b/L_b)_j}{\sum (EI/L)_j} \qquad j = A, B \qquad [3.27]$$

The story with the greatest story stability index α_{ss} can conservatively be taken as the sway stability index of the total system. This may represent a good prediction for stories separated by relative stiff beams (floors), and therefore with little interaction between columns on different stories (typically with κ_θ factors greater than 6–7.5, or degree of fixity factors R between 0.7 and 0.75 (equation [3.19] with $c = 2.4$)).

However, in cases with strong interaction between columns in adjacent stories (relatively flexible beams and stiff columns), that approach may give very conservative estimates. An improved approximation of system instability may in such cases be obtained by applying the MOM to the story stability indices defined above (by pseudo-critical loads). Thus:

$$\alpha_{\text{system}} \approx \bar{\alpha}_{ss} = \frac{1}{n} \sum_{}^{n} \alpha_{ss,i} \qquad [3.72]$$

where the summation should be taken over a limited number n of consecutive stories such as to reflect the interaction between the stories and to provide a maximum local mean. Normally, it should be taken over the story with the largest story stability index and the adjacent story above and below, and as many additional consecutive stories to either side that are necessary in order to obtain the maximum local mean value of the region. For low rise frames, it is appropriate to include all stories in the calculation of the mean.

To make the approach less sensitive to the number of adjacent stories to include in the summation, the approach may be generalized as described in [HEL 98] by imposing the constraint:

$$\bar{\alpha}_{ss} \geq \lim \alpha_{ss,k} \qquad\qquad [3.73]$$

where $\lim \alpha_{ss,k}$ is a lower limit based on the most flexible story, here labeled k. A limiting value based on an arbitrary story i can be calculated by assuming that the columns in the story below $(i-1)$ and in the story above $(i+1)$ are flexural members, with first-order rotational stiffness coefficients (b) corresponding to the "flexural" columns' far end support conditions. See section 3.2.2, step 4 (unbraced case), for additional details. The resulting stability index limit based on story i is denoted by $\lim \alpha_{ss,i}$. The story giving the largest limit is labeled k. Thus:

$$\lim \alpha_{ss,k} = \max \left[\lim \alpha_{ss,1}, \lim \alpha_{ss,2}, ...\right] \qquad\qquad [3.74]$$

3.3.6.2. nstability prediction based on α_{ss} defined with (H/Δ_0)

It should be noted that the story stability index $\alpha_{ss,i}$ computed for story i as defined (equation [3.71] in conjunction with restraints computed with equation [3.27]) will not give the same results as the stability index defined by equation [3.48]. The reason for this is that the vertical interaction reflected by equation [3.27] is not the same as that resulting from a first-order frame analysis for $(H/\Delta_0)_i$ of a story i.

If results for $(H/\Delta_0)_i$ of the stories are available, the maximum $\alpha_{ss,i}$ based on these will normally give a representative overall stability index. Thus:

$$\alpha_{\text{system}} \approx \max(\alpha_{ss,i}) \qquad\qquad [3.75]$$

Provided the lateral load pattern consisted of concentrated horizontal loads applied at each story level, and taken as some proportion (the same at each level) of the vertical loads on that level, Horne [HOR 75] found that this estimate would be on the safe side and always within 20% of the correct result (effective lengths within 10%). The frames considered by Horne had prismatic members, and flexibility factors of $\gamma = 1/0.9 = 1.11$ for all columns. With $\gamma = 1.0$, the maximum error was reduced from 20% to about 10%, but the estimate may not longer be on the safe side.

3.3.6.3. *Application of method of means to 14-story frame*

The symmetrical 14-story, three-bay frame in Figure 3.14 was considered in [HEL 98]. It is subjected to the same symmetrical, vertical loading at each story level. At every second story level, column stiffnesses are decreased. Stiffnesses in the first story are about four times those in the top story. Exterior columns are about 1.3 times stiffer than interior columns. Beam stiffnesses, generally smaller in the middle bay than in the exterior bays, are also decreased towards the top. Full frame details and results of an exact system stability analysis are given by Kuhn [KUH 76].

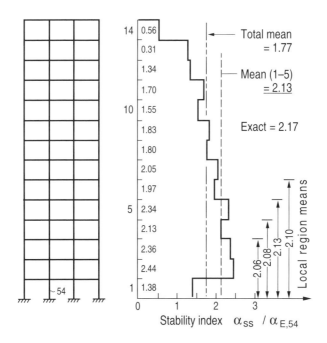

Figure 3.14. *Relative story stability indices $(\alpha_{ss}/\alpha_{E54})$ for a 14-storey, 3-bay frame*

Effective lengths, and pseudo-critical free-sway loads, N_{cs}, are calculated for κ_θ factors with $b = 6$ for the beams, corresponding to antisymmetric bending. At the exterior joints, they vary between about 1.2 at the first floor level and 2.0 at the seventh level. At the corresponding interior joints, they vary between about 2.9 and 3.8, respectively. With such κ_θ factors, the frame can probably not be classified as a typical "stiff column, flexible beam frame", but it is close to such a frame.

The story stability indices $\alpha_{ss,i}$, obtained from equation [3.71] with $\gamma_n = \gamma_s = 1$ for each story i, are given in Figure 3.14 in terms of α_E for column 54 (interior, first story) by the stepped lines. The largest value (2.44) is found for story 2.

Adopting the MOM approach, a maximum local mean for the region is obtained, as illustrated in the figure, as the mean of stories $1 - 5$. Thus:

$$\alpha_{\text{system}} \approx \bar{\alpha}_{ss}(1-5) = 2.13\,\alpha_{E\,54}$$

which is 2% below the "exact" value (2.17) obtained from the computer solution given by Kuhn. Corresponding β_0 factors become 1% lower than exact values.

3.3.6.4. *Application of MOM to 24-story frame*

Another frame considered in [HEL 98] for instability by the MOM is a symmetrical 24-story, one-bay frame. Figure 3.15 shows the lower 14 stories. The same vertical loading is applied at all story levels. The column stiffnesses are identical within sets of eight stories, but significantly different for the three sets. The columns of the lower stories are considerably stiffer than the beams: EI/L of columns are about 15 times EI_b/L_b of the beams. Thus, the frame represents a rather extreme example of a "stiff column-flexible beam frame".

Two sets of story stability indices are given in the figure. Those labeled $\alpha_{ss,\Delta}$ are based on equation [3.48] with γ_n taken as constants equal to 1.2 in the bottom story and as 1.05 in the others.

The story stability indices denoted $\alpha_{ss,c}$ (index c for critical load based) are based on the simplified version of equation [3.49] defined by equation [3.71], with $\gamma_n = \gamma_s = 1$.

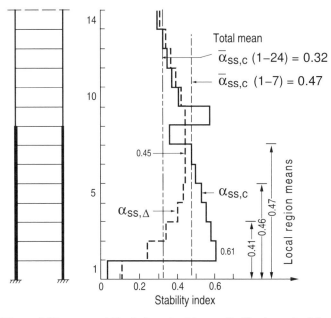

Figure 3.15. *story stability indices for 24-story "stiff column-flexible beam" frame*

The first-order bending moment diagram (due to the lateral load at the top) shows single curvature bending in the bottom four stories. The significant differences in the values and variations of $\alpha_{ss,\Delta}$ and $\alpha_{ss,c}$ in the lower stories are typical for regions with single curvature bending. The isolated conventional β_0 factors used for computing the critical free-sway loads in equation [3.71] do not reflect the strong restraint provided by the bottom story (fixed at the base) to the subsequent stories. As a consequence, the story stability indices $\alpha_{ss,c}$ may be significantly underestimated in the bottom story and overestimated in a number of subsequent stories. The first-order restraints reflected in $\alpha_{ss,\Delta}$ are in better accordance with the correct interaction, but they are still not adequate (as will be discussed further in conjunction with Figure 3.18).

Adopting the MOM approach for the $\alpha_{ss,c}$ results, a maximum local mean for the region is obtained, as illustrated in the figure, as the mean of stories $1-7$. Then:

$$\alpha_{\text{system}} \approx \bar{\alpha}_{ss,c}(1-7) = 0.47$$

For the $\alpha_{ss,\Delta}$ results, a representative instability prediction is obtained according to equation [3.75] by:

$$\alpha_{\text{system}} \approx \max(\alpha_{ss,\Delta}) = 0.45$$

The lateral loading (a single load at the top) deviates from the load pattern recommended in [HOR 75] for instability analysis, but it is believed to be reasonably acceptable. The value of 0.45 also corresponds well with 0.47 given above by the MOM. Concerning the latter, the incorrect vertical interaction reflected in the $\alpha_{ss,c}$ predictions seems to be reasonably well "corrected for" by the MOM also in the case of this rather "extreme" frame.

3.4. Story sway and moment magnification

3.4.1. *General*

3.4.1.1. *Story sway magnification*

In the previous section, approximate methods for the computation of system critical loads were considered for unbraced or partially braced single-level and multistory frames. In practical design, global second-order effects on displacements and moments are often of more importance. In this section, sway and moments in such frames will be dealt with using the system (story) stability indices already discussed (section 3.3).

The sway magnifier (given previously by equations [3.47], [3.48] and [3.49], and repeated here for convenience) can be defined by:

$$B_s = \frac{\Delta}{\Delta_0} = \frac{1}{1 - \alpha_{ss}} \qquad [3.76]$$

where

$$\alpha_{ss} = \frac{\sum(\gamma_n N/L)}{(H/\Delta_0)} \qquad [3.77]$$

or

$$\alpha_{ss} = \frac{\sum(\gamma_n N/L)}{\sum(\gamma_s N_{cs}/L) + S_B} \qquad [3.78]$$

When applied to a story of a multistory structure, H is the total shear transmitted through the story (sum of column shears), Δ_0 is the corresponding lateral displacement between the bottom and top of the story and L is the story height. For regular, multistory unbraced structures, γ_n may be approximated by γ_s.

The B_s prediction for a story in a multistory frame based on α_{ss} according to equation [3.77] will normally not be the same as the prediction for the same story based on equation [3.78]. As already discussed (section 3.3.6), the reason for this is that the vertical interaction reflected by equation [3.27] is not the same as that resulting from a total frame first-order analysis for $(H/\Delta_0)_i$ of a story i.

Often, the difference is small, such as for frames with floor beams that are reasonably stiff compared to the column stiffnesses. For "stiff column-flexible beam" stories, the difference can be significant. An example of such a case is considered in section 3.4.4.

3.4.1.2. Moment magnification

The story sway magnifier is normally also used as a moment magnifier for first-order column end moments due to lateral sway loading. As discussed in Chapter 2, section 2.4.3, and defined there and later (e.g. equations [2.39, 2.107]), end moments in the general case can be given by $B_{ei}M_{0j}^*$ and simplified to:

$$M_{0j}^* = (M_{0b} + B_s M_{0s})_j \qquad j = 1, 2 \qquad\qquad [3.79]$$

The maximum moment (Chapter 2, equations [2.14], [2.38], [2.105a]) can be given by:

$$\begin{aligned} M_{max} &= B_m\, M_{02}^* \\ &= B_m\, (M_{0b} + B_s M_{0s})_2 \end{aligned} \qquad\qquad [3.80]$$

where $M_{02}^* = (M_{0b} + B_s M_{0s})_2$ is the larger end moment (at end 2, per definition). In this formulation, B_s accounts for global (system) second-order effects and B_m for local (member) second-order effects. B_m is normally taken simplified according to equation [2.105b] or [2.106b], and greater than

1.0. In frames with sway, maximum moments will most often be at an end $(B_m = 1)$.

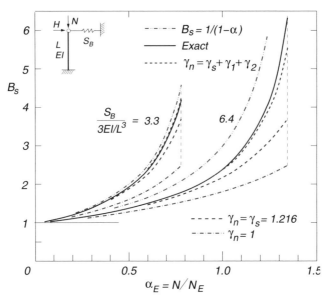

Figure 3.16. *Exact and approximate sway magnifiers of a partially braced column versus axial load level for two lateral bracing stiffnesses*

3.4.2. *Partially braced column – example*

3.4.2.1. *Sway magnifier prediction*

The partially restrained column analyzed for instability in section 3.3.4 will be considered. The system consists of a partially braced cantilever column, fully fixed at the base and pinned at the top. Results for this case are more sensitive to approximations in local, second-order member effects $(N\delta)$ than similar cases with smaller differences in rotational end restraints. It is therefore a good case for testing the applicability of the approximate methods.

The column parameters are the same as in section 3.3.4: $\beta_{0s} = 2$, $\beta_{0b} = 0.7$, $\alpha_{s,b} = (2/0.7)^2 = 8.163$, $\alpha_s = N/N_{cs} = \alpha_E \beta_{0s}^2 = 4\alpha_E$.

Typical sway magnifier results selected from [HEL 09a] are shown versus the nominal load index $\alpha_E = N/N_E$ in Figure 3.16 for the system shown by the insert in the same figure. The exact solutions (full lines) are obtained using standard stiffness relationships defined with stability functions.

The two sets of curves are for the relative bracing stiffnesses $S_B/(3EI/L^3) = 3.3$ and 6.4, corresponding to a moderate and relative strong lateral restraint. Critical load indices (to one decimal), at system instability, for these two cases are $\alpha_{E,cr} = N_{cr,exact}/N_E = 1.0$ and 1.5, respectively. As these values are approached, sway deflections, and B_s, tend toward infinity. This corresponds, respectively, to four and six times the free-sway load index of $\alpha_{E,s} = N_{cs}/N_E = 0.25$ $(\beta_s = 2)$ and to about 0.49 and 0.74 times the braced load index of $\alpha_{E,b} = N_{cb}/N_E = 2.04$. The curves are arbitrarily terminated at the load level at, which, the results for the case $\gamma_n = 1$ give $B_s = 2.5$.

The approximate B_s results in the figure have been obtained for the three γ_n assumptions defined by:

1) $\gamma_n = 1$

2) $\gamma_n = \gamma_s = 1.216$

3) $\gamma_n = \gamma_s + \gamma_1 + \gamma_2 = 1.216 + 0.026\,(4\alpha_E - 1) + 4.90\,[(4\alpha_E - 1)/8.16]^8$

In addition, results are presented for:

$$B_s = \frac{1}{1-\alpha} \quad \text{where} \quad \alpha = \frac{N}{N_{cr,exact}} \qquad [3.81]$$

defined with the exact system critical load (given by the full line in Figure 3.11). Predictions by equation [3.81] are conservative, and become increasingly so for increasing lateral restraint.

Results obtained with the extended $\gamma_n = \gamma_s + \gamma_1 + \gamma_2$ are seen to be in very good agreement with exact results at any load level and for any lateral restraint. The other approximate results deviate to an increasing extent from the exact results with increasing load level and increasing lateral restraint.

The use of $\gamma_n = 1$ underestimates B_s, and is not recommended, possibly except at very low load levels. For $\gamma_n = \gamma_s = 1.216$, a good agreement with exact results is obtained in the practical range for B_s less than about 1.5. For stronger lateral restraints than those shown here, the underestimation of B_s may become unacceptable also for $\gamma_n = \gamma_s = 1.216$ [HEL 96a].

3.4.2.2. *Moments*

The larger end moment is at the base (end 2). The top is pinned with $M = 0$. The maximum moment, in this case with lateral load only, is given by:

$$M_{max} = B_m \, B_s M_{0s,2}$$

The approximate maximum magnifier $B_m \, (\geq 1)$ given in Chapter 2 by equation [2.105b] or [2.106b] will, for $C_m = 0.6$ and $N_B = N_E/0.7^2$, become greater than 1.0 for $\alpha_E > 0.72$ and 0.82, respectively. For practical sway magnifiers limited to about 1.5, this means that the column with the stronger lateral restraint will be affected by $B_m \, (\geq 1)$ for $\alpha_E > 0.72$, or 0.82. For the column with the weaker lateral restraint, however, the maximum moment will always be at the end and given by $M_{max} = B_s M_{0s2}$.

From results of an exact analysis, it can be found that the exact maximum moment magnifier B_m is less than 1.0 for α_E less than about 0.93, and that the approximate moment predictions are quite conservative for the considered restrained fixed-pinned column.

3.4.3. *Partially braced frame – example*

3.4.3.1. *Sway magnifier prediction*

As an example of a sway magnifier calculation for a frame, the three-bay frame in Figure 3.12 is considered once more. Frame data and applied load distribution are the same as given in section 3.3.5.

Making use of equations [3.69] and [3.70], the sway magnifier becomes:

$$\alpha_{ss} = \frac{\sum \gamma_n N/L}{H/\Delta_0} = \frac{\left(0.5\gamma_{n1} + \gamma_{n2} + \dfrac{2\gamma_{n3}}{1.5}\right) P}{0.501 \, N_E} \tag{3.82}$$

The sway magnifier will be calculated for three γ_n cases, two of which have been considered in section 3.3.5. They are defined by:

Case 1. $\gamma_n = 1$: $\alpha_{ss} = 5.655 \frac{P}{N_E}$

Case 2. $\gamma_n = \gamma_s$: $\alpha_{ss} = 6.445 \frac{P}{N_E}$

Case 3. $\gamma_n = \gamma_s + \gamma_1$: $\alpha_{ss} = (6.351 + 2.597 \frac{P}{N_E}) \frac{P}{N_E}$

Sway magnifier predictions for these three cases are shown in Figure 3.17. For $\gamma_n = \gamma_s + \gamma_1$, the agreement with exact results is excellent. This is also the case for results obtained with $\gamma_n = \gamma_s$. For practical values (less than about $B_s = 1.5$), there is almost no difference. The simplest assumption, $\gamma_n = 1$, is again not so good.

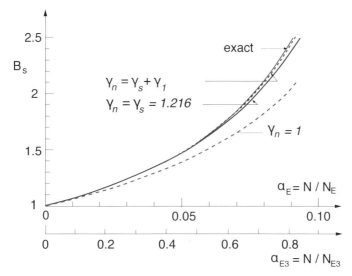

Figure 3.17. *Exact and approximate sway magnifiers versus axial load level for the three-bay frame in Figure 3.12*

3.4.3.2. *Moments*

For columns 1 and 4, maximum moments will always be at the end, and given by $M_{max} = B_s M_{0s,2}$. Also for the very flexible column 3, maximum moments will be at the end for practical, and even considerably higher, sway magnifiers (for $\alpha_{E3} < 0.72$ and 0.82 when B_m is taken according to equations [2.105b, 2.106b], respectively).

3.4.4. *Sway magnifier prediction of frames with single curvature regions*

3.4.4.1. *Sway magnifier*

As mentioned in section 3.3.1, the approximate second-order story magnifier method, with the defined γ_s and γ_n factors, may significantly underestimate sway and moments of stories located in single curvature regions of unbraced multistory frames. Such single curvature regions may result in frames with stiff columns and flexible beams. In such cases, there may be significant interaction between columns of adjacent stories, and this interaction is not properly reflected by first-order column end restraints implied in equation [3.48], and even less so in equation [3.49] that is based on simplified vertical interaction assumptions (equation [3.33]).

These local second-order effects on rotational restraints are complicated and not well understood. From a study of typical story sway magnification factors in the negative curvature regions of multi-story frames [HEL 09b], it was found that these local second-order effects could be approximately accounted for through a reasonably simple "restraint correction flexibility factor" expression denoted γ_k.

In conjunction with the story stability index given in terms of the first-order stiffness (H/Δ_0) by equation [3.77], it was proposed to compute the story sway magnifier for each story from equation [3.76]:

$$B_s = \frac{1}{1 - \alpha_{ss}}$$

but with γ_n in the sway stability index expression (equation [3.77]) replaced by a *combined* flexibility factor γ_c such that:

$$\alpha_{ss} = \frac{\sum(\gamma_c N/L)}{H/\Delta_0} \tag{3.83}$$

where

$$\gamma_c = \gamma_n \gamma_k \approx \gamma_s \gamma_k \tag{3.84}$$

and

$$\gamma_k = (0.11\frac{L_{sc}}{x_{ts}} + 0.89)^3 \geq 1 \qquad [3.85]$$

The flexibility factors γ_n and γ_k account for second-order effects of axial forces acting on the bent shape ($N\delta$ effects) and on the rotational restraints (vertical interaction effects), respectively. L_{sc} is the length of the single curvature region, taken as the distance along the structure from the base to the first-order inflection point (at the top of the single curvature region), and x_{ts} is the distance from the base of the structure to the top of the story considered.

The constants in the γ_k expression are dependent on several factors, including axial load levels. The constants adopted in equation [3.85] are chosen such as to provide acceptable predictions for common frames in the practical magnifier range of about $B_s = 1.3 - 1.5$.

The α_{ss} formulation above is valid for frames both with and without single curvature regions. For frames with columns in double curvature, and for columns on stories above single curvature regions, equation [3.85] gives $\gamma_k = 1$ since $x_{ts} > L_{sc}$ in such cases. The γ_n factor may be approximated by γ_s in most practical applications. This is, in particular, the case for columns in single curvature regions.

Columns located along one vertical axis of a multi-story and multi-bay frame may have different γ_s and γ_k factors from those of columns located along another vertical axis of the frame. This may be so due to different L_{sc} values, caused by differences in restraints, in the various axes. The sway is dependent on the combined effect from all axes and is reflected through the summation in equation [3.83].

3.4.4.2. *Application to 24-story "stiff column-flexible beam" frame*

Results of an application of the approach to a multi-story "stiff column-flexible beam" frame are shown in Figure 3.18 (from [HEL 09b]). The same frame was considered before in Figure 3.15 for system instability and described in that connection.

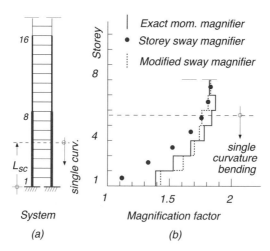

Figure 3.18. *Magnifiers for 24-story "stiff column-flexible beam" frame*

The story sway magnifiers B_s marked with dots in Figure 3.18 are taken from [LAI 83a]. So are the exact moment magnifiers (solid stepped lines) for column end moments (M_2/M_{02}). The story sway magnifiers are based on stability indices given by equation [3.48] with constant flexibility factors of γ equal to 1.2 in the bottom story and 1.05 in the others. These correspond to those used for results denoted by "$\alpha_{ss,\Delta}$" in Figure 3.15.

Modified sway magnifier predictions, B_s, with stability indices given by equation [3.83] are also shown in the figure (dashed, stepped lines). They were calculated with a length of the single curvature region of $L_{sc} = 4.7L$, where L is the storey height, and ratios L_{sc}/x_{ts} of $4.7L/L = 4.7$ for story 1, $4.7L/2L = 2.35$ for story 2, $4.7L/3L = 1.57$ for story 3, $4.7L/4L = 1.18$ for story 4, and $4.7L/5L = 0.94$ (<1) for story 5. The corresponding γ_k values (equation [3.85]) are 2.79, 1.52, 1.20, 1.06 and 1.0. The γ_n factors are taken equal to those used in [LAI 83a] (1.2 in bottom story and 1.05 for the others).

Maximum moments are at column ends. Thus, in the approximate method, the sway magnifier is also the moment magnifier ($M_{max} = B_m B_s M_{02} = B_s M_{02}$). The modified story sway magnifier results are seen to be in good agreement with the exact moment magnifiers (between 5% above and 2% below) for this frame.

3.4.5. *Iterative elastic analysis method*

There are a number of other approximate methods available in the literature. These include iterative methods in which the effect of axial forces on displacements is replaced by equivalent lateral loads in several iterations in order to converge toward a final state. The method is often referred to as $P - \Delta$ or $N - \Delta$ analysis. The major steps in the method are described with reference to the multi-story frame in Figure 3.19.

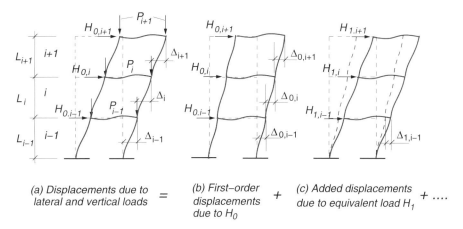

(a) Displacements due to   =   (b) First–order   +   (c) Added displacements   + ...
lateral and vertical loads   displacements   due to equivalent load H_1
                                   due to H_0

Figure 3.19. *Iterative elastic analysis method*

Figure 3.19(a) shows the frame in its final displaced state. The relative displacements between the bottom and top of a story i is labeled Δ_i and is due to the effect of given lateral loads $H_{0,i}$ (including effects of imperfection inclinations) and total vertical loading at each floor level denoted by P_i.

Since the final displacements are not known, the analysis starts off by assuming a first approximation to be given by the relative first-order displacements $\Delta_{0,i}$, Figure 3.19(b). The overturning effect of the vertical loads times the relative displacements is accounted by a fictitious horizontal ("shear") load at each floor (top of each storey). For story i, it can be defined by:

$$H_{1,i} = \frac{\bar{\gamma}_i P_i \Delta_{0,i}}{L_i} \qquad\qquad [3.86]$$

where $\bar{\gamma}_n$ is an average flexibility factor for the columns in story i, which accounts for the increased flexibility of columns with nonlinear moment distributions (see previous discussion in section 3.3.2).

When applying these global sway loads to the top of each story, it results in additional relative displacements $\Delta_{1,i}$, Figure 3.19(c), which in turn gives rise to new fictitious global sway loads:

$$H_{2,i} = \frac{\bar{\gamma}_i P_i \Delta_{1,i}}{L_i} \qquad\qquad [3.87]$$

Generally, one or two iterations are satisfactory for most frames.

Early work on this method is reviewed in [LAI 83a]. In early presentations of the method, and also mostly since, $\bar{\gamma}_n$ is not included. Later presentations (e.g. [LAI 83a]) included such effects. With $\bar{\gamma}_n$ included, the method as presented here is strictly a modified iterative method.

Clearly, in an analysis with lateral forces, real or fictitious, shear forces result. However, it is only the shear forces caused by the real lateral forces that should be included in possible design contexts. Those due to the fictitious global sway forces must be deducted from the total ones.

It is somewhat cumbersome to carry out successive first-order analyses. It would be simpler if we could determine the total, global sway modified lateral load in order to carry out only one first-order analysis in addition to the initial one (for the applied real lateral loads $H_{0,i}$) An extension obtained by developing a closed-form solution is presented in section 3.4.6 (global sway magnified loading based on an iterative approach).

3.4.6. Global magnifiers for sway and moments

3.4.6.1. Single global magnifier based on system critical load

In Chapter 2, section 2.4.4, the well-known sway magnifier defined by equation [2.15] was applied to a single member. A similar magnifier (equation [3.81]) was included as an alternative in the partially braced column example (Figure 3.16).

If applying it to a multistory frame, it can be written as:

$$B_s = \frac{1}{1-\alpha} \quad \text{with} \quad \alpha = \frac{(\sum N)_{base}}{N_{cr}} \qquad [3.88]$$

where $(\sum N)_{base}$ is the total vertical load acting on the structure, which is equal to the sum of axial forces of all members (braced or bracing) in the bottom (base) storey, and N_{cr} is the system critical load of the multi-story frame, including effects of possible shear walls, other bracing elements, etc.

This approach, whereby moments in the structure are multiplied by the same multiplier, is very simple, but not surprisingly, very conservative for large portions of the structure.

Resulting moments can be denoted as global sway modified first-order moments. They can, as before, be written as:

$$M_0^* = B_s M_{0s}$$

where M_{0s} is the first-order moment due to the real lateral loads (including imperfection effects). If combined with other loadings, the maximum moment in each column can again be given by equation [3.80].

As an alternative to multiplying first-order end moments by B_s after the first-order analysis is carried out for an applied lateral loading "$F_{H,0}$", we could of course run the first-order analysis for the magnified lateral loading "$B_s F_{H,0}$". The net result would be the same.

The latter approach has been adopted by EC2:2004 in Annex H.2. With the EC2 notation, the magnified lateral loading to be applied to the frame is defined by:

$$F_H = \frac{F_{H,0}}{1 - F_V/F_{V,B}} \qquad [3.89]$$

where $F_{H,0}$ is the horizontal force load due to wind, imperfections, etc., somewhat strangely called "first-order" horizontal force; F_V is the total vertical load on bracing and braced members (equal to $(\sum N)_{base}$ above); and $F_{V,B}$ is the global buckling load (equal to N_{cr} above). The subscript Ed used by EC2 to reflect the ultimate limit state, is for simplicity omitted in the notation above.

Although the resulting moments due to the magnified lateral loading are obtained from a first-order analysis, it would be incorrect to call them first-order moments because they include global second-order effects through the magnification factor. EC2 is not clear on this point. For clarity, moments due to the global sway modified (magnified) horizontal load F_H may as before be named global sway modified first-order moments and labeled as M_0^* (with a star super index), as adopted elsewhere in this book.

Finally, it should be recalled that correct shear forces in columns, shear walls, etc., are not due to the magnified lateral load F_H, but due to the real applied lateral loads, such as wind. Equivalent lateral loads to account for imperfection inclinations should strictly not be included when calculating shear forces. The distribution of shear forces to the different columns in a story of a multi-bay frame will be affected by local second-order effects, but the sum of shears (and possible bracing forces) will always be equal to the sum due to the applied lateral loads only.

Application: In the instability evaluation of the 24-storey frame (Figure 3.15), it was found that the system stability index was about α_{ss} = 0.45 to 0.47. If assuming the average value α_{ss} = 0.46, the system sway magnifier (equation [3.88]) becomes:

$$B_s = \frac{1}{1 - 0.46} = 1.85.$$

It can be seen in Figure 3.18 that this factor will be adequate for moments in all stories, but quite conservative both in the lower and upper regions of the frame. At a preliminary design stage, such an overall global factor is useful. For final design, values at each story level would be of interest.

3.4.6.2. *Global sway magnifier based on the bottom (base) storey*

In cases when global buckling loads, $F_{V,B}$, are not defined, EC2:2004 Annex H.2 suggests that the magnified lateral load F_H may be taken according to:

$$F_H = \frac{F_{H,0}}{1 - F_{H,1}/F_{H,0}} \qquad [3.90]$$

where $F_{H,0}$ is defined above and $F_{H,1}$ is a fictitious horizontal load that gives the same bending moment as the vertical load N_V acting on the lateral displacement caused by $F_{H,0}$.

It is not immediately apparent how this expression is intended to be applied, partly because N_V is simply defined as a vertical load. This is not a very clear definition. One interpretation is that equation [3.90] should be used to obtain a single multiplier for the whole frame, for instance, based on the bottom (base) story, with N_V equal to total vertical loads, denoted by F_V above, and with the equivalent horizontal force at the top of the story (of height L) expressed by $F_{H,1} = F_V \Delta_0 / L$. Then, equation [3.90] for the story can be rewritten in terms as $F_H = B_s F_{H,0}$, where the sway magnifier is given by:

$$B_s = \frac{1}{1 - \alpha_{ss}} \quad \text{and} \quad \alpha_{ss} = \frac{(F_V/L)}{(F_{H,0}/\Delta_0)} = \frac{(\sum N/L)_{base}}{(H/\Delta_0)_{base}} \qquad [3.91]$$

Apart from the missing γ_n factor, this is the same story sway magnifier that can be obtained with α_{ss} given by equation [3.77].

In Figure 3.18, it can be seen that the magnifier for the first (bottom) story is not the most unfavorable magnifier. Indeed, the sway magnifier for the bottom story can be seen to seriously underestimate the maximum moment magnifier. Fixity, or near fixity, imparts a stiffness to the story and adjacent stories that more than balances the greater vertical and horizontal loads the bottom storey has to carry. So, the magnifier for this story is not acceptable for computing a single global magnification formulation. For other, less extreme frames, it may provide a better estimate than for the example considered, but it is not recommended as a general approach.

3.4.6.3. *Global sway magnified loading based on an iterative approach*

Probably, another manner of application of equation [3.90] than the one above is intended. In a note following the presentation of the expression, mention is made that it follows from a "step-by-step numerical calculation". This points to the type of iterative method discussed in section 3.4.5 in conjunction with Figure 3.19.

As mentioned therein, the iterative approach would be simplified if one could determine a total, global sway modified lateral load at the different levels (stories), in order to avoid having to carry out more than one first-order analysis in addition to the initial one (for the applied real lateral loads $H_{0,i}$).

Still with reference to Figure 3.19, it would seem reasonable to assume that the proportionality relationships:

$$\Delta_{1,i}/\Delta_{0,i} = H_{1,i}/H_{0,i} \; ; \; \Delta_{2,i}/\Delta_{0,i} = H_{2,i}/H_{0,i} \; ; \; \text{etc.}$$

may be reasonably valid for regular structures and loadings. Then, for such cases, the total lateral load, including fictitious global sway forces, can be written:

$$
\begin{aligned}
H_i &= H_{0,i} + H_{1,i} + H_{2,i} + \cdots \\
&= H_{0,i} \left[1 + \frac{H_{1,i}}{H_{0,i}} + \frac{H_{2,i}}{H_{0,i}} + \cdots \right] \\
&= H_{0,i} \left[1 + \alpha_i + \alpha_i^2 + \alpha_i^3 + \cdots \right] \\
&= H_{0,i} \, \frac{1}{1 - \alpha_i} \quad (\text{for} \quad \alpha_i < 1)
\end{aligned}
\qquad [3.92a]
$$

where

$$\alpha_i = \frac{H_{1,i}}{H_{0,i}} \qquad\qquad [3.92b]$$

or

$$\alpha_i = \frac{\bar{\gamma}_i P_i \Delta_{0,i}}{H_{0,i} L_i} \qquad\qquad [3.93]$$

when substituting for H_1 from equation [3.86]. Here, P_i is the total vertical load on the floor level above storey i. This load will be equal to the difference in axial column forces in story i and $i+1$ ($P_i = \sum N_i - \sum N_{i+1}$).

The results obtained from a first-order analysis with the magnified lateral load H_i applied at each floor level i include global second-order sway effects through the global sway magnifier $(1/(1 - \alpha_i))$ at each floor level. The resulting moments are consequently global sway modified first-order moments M_0^*.

By replacing H above by F_H, H_0 by $F_{H,0}$ and H_1 by $F_{H,1}$, the final form of equation [3.92] can be rewritten in the form of the EC2 equation, equation [3.90]. The EC2 formulation does not include the flexibility factor, and it does not give the detailed form of H_1/H_0 defined by equation [3.93]. From the above equation it should be clear that the EC2 equation, equation [3.90], does not imply a single magnifier, but rather a separate magnifier for each storey.

As mentioned, the EC2 formulation is vague and unclear. Some more details, and a discussion of various aspects of the approach, are given in [WES 04], but also therein the presentation is of a rather general form that leaves some questions unanswered.

3.4.6.4. *Other approximate global second-order methods*

There is a number of other approximate methods available in the literature. These include the "overturning moment magnifier method" and the "frame magnifier method". Both of these aim at providing a single global sway magnifier and are possibly of main use in frames with shear walls. For a brief review of these, see [LAI 83a].

Appendix 1

Cardano's Method

A1.1. Introduction

This appendix gives the mathematical method to solve the roots of a polynomial of degree three, called a cubic equation. Some results in this section can be found, for instance, in [ART 04].

As a useful extension, we also give the methodology to determine the roots of a polynomial of degree four, called a quartic equation. The roots of a cubic equation, like those of a quartic equation, can be found algebraically. It can be shown that this property is no more valid, in general, for a quintic equation (equation of fifth-order) or equations of higher degrees. This is known as the Abel–Ruffini theorem, which was first published in 1824.

Referring to the French dictionary *Le Robert*, the complete method for the general resolution of a cubic is probably due to Tartaglia (Niccolo Fontana, 1500–1557, also called Tartaglia) from his works concluded in 1537, based on the first approach of Gerolamo Cardano (1501–1576). In 1539, Tartaglia revealed his method to Cardano on the condition that Cardano would never reveal it. Some years later, Cardano learned about Ferro's prior work and published Ferro's method in his book *Ars magnasive de regulis algebraicis, liner unus* in 1545. These works, which are produced by the Italian mathematics school, are also based on:

– Rafael Bombelli (1526–1572) was the one who finally managed to address the resolution of polynomial equations with imaginary numbers.

– Lodovico Ferrari (1522–1565), as Cardano's student, gave the solution of a quartic equation, which was published in one chapter of *Ars magnasive de regulis algebraicis, liner unus* written by Cardano in 1545.

– Scipione del Ferro (1465–1526) first discovered the method to solve the canonical form of a cubic equation ($x^3 + px + q = 0$), the first step toward the more general method of a cubic equation.

In the following, we use the mathematical function sgn(x) for the sign function of a real x, and we also use:

$$\sqrt[3]{x} = |x|^{1/3}.sgn(x) \text{ with } |x|=sgn(x).x \qquad [\text{A1.1}]$$

A1.2. Roots of a cubic function – method of resolution

A1.2.1. *Canonical form*

We consider the cubic equation with real coefficients:

$$g(t) = at^3 + bt^2 + ct + d = 0 \text{ with } a \neq 0 \qquad [\text{A1.2}]$$

Each term can be divided by the first coefficient, leading to:

$$t^3 + \frac{b}{a} t^2 + \frac{c}{a} t + \frac{d}{a} = 0 \qquad [\text{A1.3}]$$

This cubic equation can be factorized as:

$$(t + \frac{b}{3a})^3 + \frac{ac - b^2}{3a^2}(t + \frac{b}{3a}) + \frac{27a^2d + 2b^3 - 9abc}{27a^3} = 0 \qquad [\text{A1.4}]$$

which is known as the canonical form:

$$f(x) = x^3 + px + q = 0 \text{ by setting}$$
$$x = t + \frac{b}{3a} \;; p = \frac{3ac - b^2}{3a^2} \text{ and } q = \frac{27a^2d + 2b^3 - 9abc}{27a^3} \qquad [\text{A1.5}]$$

This canonical equation is solved from the introduction of two numbers y and z such that $x = y + z$ are roots of the cubic equation $f(x) = 0$, by imposing some constraint equalities:

$$\begin{cases} y^3 + z^3 = -q \\ yz = -\dfrac{p}{3} \end{cases} \Leftrightarrow \begin{cases} f(y + z) = 0 \\ y^3 + z^3 = -q \\ yz = -\dfrac{p}{3} \end{cases} \Leftrightarrow \begin{cases} (y^3 + z^3) + (p + 3yz)(y + z) + q = 0 \\ y^3 + z^3 = -q \\ yz = -\dfrac{p}{3} \end{cases}$$

$$[\text{A1.6}]$$

The following change of variable can be chosen as:

$$\begin{cases} Y = y^3 \\ Z = z^3 \end{cases} \text{ and then } \begin{cases} Y + Z = -q \\ (YZ)^{1/3} = -\dfrac{p}{3} \end{cases} \qquad [\text{A1.7}]$$

Knowing the sum and the product of Y and Z, these numbers are necessarily the roots U_1 and U_2 of the quadratic equation: $U^2 + qU - \dfrac{p^3}{27} = 0$. If U_1 and U_2 are known, then y and z are calculated from $(e^{2kip} U_1)^{1/3}$ and $(e^{2kip} U_2)^{1/3}$, which should be associated by a pair such that the product yz is a real number. We can distinguish several possible cases using the discriminant concept, depending on the sign of $D = q^2 + \dfrac{4p^3}{27}$ or equivalently, depending on the sign of $4p^3 + 27q^2$.

A1.2.2. Resolution – one real and two complex roots

Case $4p^3 + 27q^2 > 0$ (one real and two complex conjugate roots for $f(x) = 0$).

This case includes the case $p = 0$.

In this case, both U_1 and U_2 are real numbers:

$$U_1 = \frac{-q + \sqrt{q^2 + \dfrac{4p^3}{27}}}{2} = -\frac{q}{2} + \sqrt{\frac{q^2}{4} + \frac{p^3}{27}} \quad \text{and}$$

$$U_2 = -\frac{q}{2} - \sqrt{\frac{q^2}{4} + \frac{p^3}{27}} \qquad [\text{A1.8}]$$

To have the product yz as a real number, the possible couples (y, z) (or equivalently (z,y)) are then:

$$(\sqrt[3]{U_1};\sqrt[3]{U_2}),\ (j\sqrt[3]{U_1};j^2\sqrt[3]{U_2}),\ (j^2\sqrt[3]{U_1};j\sqrt[3]{U_2}) \qquad [A1.9]$$

where j denotes the complex number that is the cubic root of unity. The solutions in x are then:

$$x_1 = \sqrt[3]{-\frac{q}{2}+\sqrt{\frac{q^2}{4}+\frac{p^3}{27}}} + \sqrt[3]{-\frac{q}{2}-\sqrt{\frac{q^2}{4}+\frac{p^3}{27}}}$$

$$x_2 = j\sqrt[3]{-\frac{q}{2}+\sqrt{\frac{q^2}{4}+\frac{p^3}{27}}} + j^2\sqrt[3]{-\frac{q}{2}-\sqrt{\frac{q^2}{4}+\frac{p^3}{27}}}$$

with $j=e^{2i\pi/3}$ $\qquad [A1.10]$

$$x_3 = j^2\sqrt[3]{-\frac{q}{2}+\sqrt{\frac{q^2}{4}+\frac{p^3}{27}}} + j\sqrt[3]{-\frac{q}{2}-\sqrt{\frac{q^2}{4}+\frac{p^3}{27}}}$$

In reinforced concrete design, we are only concerned with real solutions, and then only x_1 will be of interest, which finally leads to the root of the initial cubic equation [A1.2], as:

$$t = \frac{1}{3a}\sqrt[3]{-\frac{27a^2d+2b^3-9abc}{2}+\sqrt{\left(\frac{27a^2d+2b^3-9abc}{2}\right)^2+(3ac-b^2)^3}} +$$

$$\frac{1}{3a}\sqrt[3]{-\frac{27a^2d+2b^3-9abc}{2}-\sqrt{\left(\frac{27a^2d+2b^3-9abc}{2}\right)^2+(3ac-b^2)^3}} -\frac{b}{3a}$$

$$[A1.11]$$

In the specific case $p = 0$, this real root is simply reduced to

$$t=\frac{1}{3a}\sqrt[3]{-27a^2d-2b^3+9abc}-\frac{b}{3a}$$

A1.2.3. *Resolution – two real roots*

Case $4p^3 + 27q^2 = 0$ (one real and one double real roots for $f(x) = 0$).

In this case, U_1 and U_2 are real numbers with $U_1 = U_2 = -\dfrac{q}{2} = \left(\dfrac{3q}{2p}\right)^3 = U$. The product yz being real, the possible couples (y, z) (or equivalently (z,y)) are given by:

$$(\sqrt[3]{U} \; ; \sqrt[3]{U} \;), (j \sqrt[3]{U} \; ; j^2 \sqrt[3]{U} \;), (j^2 \sqrt[3]{U} \; ; j \sqrt[3]{U} \;) \qquad\qquad \text{[A1.12]}$$

where j denotes the complex number that is the cubic root of unity. Using the fundamental property $1 + j + j^2 = 0$, the solutions in x are given by:

$$\text{simple root: } x_1 = 2 \sqrt[3]{-\dfrac{q}{2}} = \dfrac{3q}{p} \; ;$$

$$\text{double root: } x_2 = x_3 = -\dfrac{3q}{2p} \qquad\qquad \text{[A1.13]}$$

The real roots of the initial cubic equation $g = 0$ in "t" (equation [A1.2]) are then:

$$t_1 = \dfrac{3q}{p} - \dfrac{b}{3a} = \dfrac{9a^2d + b^3 - 4abc}{a(3ac - b^2)} \quad \text{and}$$

$$t_2 = t_3 = -\dfrac{3q}{2p} - \dfrac{b}{3a} = \dfrac{-9ad + bc}{2(3ac - b^2)} \qquad\qquad \text{[A1.14]}$$

A1.2.4. *Resolution – three real roots*

Case $4p^3 + 27q^2 < 0$ (three real roots for $f(x) = 0$).

In this case, U_1 and U_2 are conjugate imaginary numbers. It can be checked that if y is a cubic root of the complex number U_1, then $z = -\dfrac{p}{3y}$ is the imaginary conjugate number of y; and $x = y+z$ is a real number. Practically, we use the fact that a necessary and sufficient condition for two polynomial equations to have the same roots is that the coefficients of these polynomial equations are proportional. We use the equality $4 \cos^3 a - 3 \cos$

$a - \cos 3a = 0$ which is always true. The unknown $y = \cos a$ is a root of $4 y^3 - 3 y - \cos 3a = 0$. We are looking for the conditions to have both equations $x^3 + px + q = 0$ and $4 y^3 - 3 y - \cos 3a = 0$ with proportional coefficients. In this case, if x is a root of the first cubic, kx would be the root of the second cubic, with k as a proportional coefficient, leading to the equivalence principle:

$$\frac{1}{4 k^3} = -\frac{p}{3 k} = -\frac{q}{\cos 3a} \quad \text{for } k \neq 0 \qquad [A1.15]$$

These two equations are equivalent to the conditions:

$$\begin{cases} k^2 = -\dfrac{3}{4p} \\ k^3 = -\dfrac{\cos 3a}{4q} \end{cases} \text{ or equivalently } \begin{cases} k = \sqrt{-\dfrac{3}{4p}} \\ k^3 = -\dfrac{\cos 3a}{4q} \end{cases} \text{ when } p < 0 \quad [A1.16]$$

The elimination of k gives: $\cos 3a = \dfrac{3q}{2p}\sqrt{-\dfrac{3}{p}}$ which should be comprised between -1 and $+1$, leading to $\left(\dfrac{3q}{2p}\sqrt{-\dfrac{3}{p}}\right)^2 \leq 1 \Leftrightarrow \dfrac{27q^2}{-4p^3} \leq 1$ with $-4p^3 > 0$. We recognize the condition that the discriminant is negative, that is $4p^3 + 27q^2 \leq 0$. In this case, and from equation [A1.16], we have the inverse relationship:

$$k = \sqrt{-\frac{3}{4p}} \text{ and } a = \frac{1}{3}\left[\text{Arc} \cos\left(\frac{3q}{2p}\sqrt{-\frac{3}{p}}\right) + 2k\pi\right] \qquad [A1.17]$$

As the roots of the cubic equation $4 y^3 - 3 y - \cos 3a = 0$ are $y = \cos a$, the roots of the cubic equation $x^3 + px + q = 0$ are $1/k$ proportional to the previous ones (with $x_1 < x_2 < x_3$):

$$x_1 = 2\sqrt{-\frac{p}{3}} \, \cos\left[\frac{\text{Arc} \cos\left(\dfrac{3q}{2p}\sqrt{-\dfrac{3}{p}}\right) + 2\pi}{3}\right]$$

$$x_2 = 2\sqrt{-\frac{p}{3}} \, \cos\left[\frac{\text{Arc} \cos\left(\dfrac{3q}{2p}\sqrt{-\dfrac{3}{p}}\right) + 4\pi}{3}\right] \qquad [A1.18]$$

$$x_3 = 2\sqrt{-\frac{p}{3}}\ \cos\left[\frac{\text{Arc cos}\left(\frac{3q}{2p}\sqrt{-\frac{3}{p}}\right)}{3}\right]$$

The roots of the initial cubic equation [A1.2] $g(t) = 0$ are then (with $t_1 < t_2 < t_3$):

$$t_1 = \frac{2}{3|a|}\sqrt{b^2 - 3ac}\ \cos\left\{\frac{\text{Arc cos}\left[\frac{\text{sgn(-a)}\ (27a^2d+2b^3-9abc)(b^2-3ac)^{-1.5}}{2}\right] + 2\pi}{3}\right\} - \frac{b}{3a}$$

$$t_2 = \frac{2}{3|a|}\sqrt{b^2 - 3ac}\ \cos\left\{\frac{\text{Arc cos}\left[\frac{\text{sgn(-a)}\ (27a^2d+2b^3-9abc)(b^2-3ac)^{-1.5}}{2}\right] + 4\pi}{3}\right\} - \frac{b}{3a}$$

$$t_3 = \frac{2}{3|a|}\sqrt{b^2 - 3ac}\ \cos\left\{\frac{\text{Arc cos}\left[\frac{\text{sgn(-a)}\ (27a^2d+2b^3-9abc)(b^2-3ac)^{-1.5}}{2}\right]}{3}\right\} - \frac{b}{3a} \quad \text{[A1.19]}$$

A1.3. Roots of a cubic function – synthesis

A1.3.1. *Summary of Cardano's method*

Considering the cubic equation now expressed in terms of the unknown α that is related to the dimensionless neutral axis position in reinforced concrete design:

$$a\alpha^3 + b\alpha^2 + c\alpha + d = 0 \quad \text{[A1.20]}$$

The parameters p and q can be introduced as:

$$p = \frac{3ac - b^2}{3a^2} \text{ and } q = \frac{27a^2d + 2b^3 - 9abc}{27a^3} \quad \text{[A1.21]}$$

If $4p^3 + 27q^2 > 0$, the unique real solution of the cubic equation is obtained from:

$$\alpha_1 = \sqrt[3]{-\frac{q}{2} + \sqrt{\frac{q^2}{4} + \frac{p^3}{27}}} + \sqrt[3]{-\frac{q}{2} - \sqrt{\frac{q^2}{4} + \frac{p^3}{27}}} - \frac{b}{3a} \quad \text{[A1.22]}$$

If $4p^3 + 27q^2 < 0$, the three real solutions are given by:

$$\left\{ \begin{array}{l} \alpha_1 = 2\sqrt{\dfrac{-p}{3}}\cos\left[\dfrac{Arc\cos\left(\dfrac{3q}{2p}\sqrt{\dfrac{3}{-p}}\right)+2\pi}{3}\right]-\dfrac{b}{3a} \\[40pt] \alpha_2 = 2\sqrt{\dfrac{-p}{3}}\cos\left[\dfrac{Arc\cos\left(\dfrac{3q}{2p}\sqrt{\dfrac{3}{-p}}\right)+4\pi}{3}\right]-\dfrac{b}{3a} \\[40pt] \alpha_3 = 2\sqrt{\dfrac{-p}{3}}\cos\left[\dfrac{Arc\cos\left(\dfrac{3q}{2p}\sqrt{\dfrac{3}{-p}}\right)}{3}\right]-\dfrac{b}{3a} \end{array} \right. \qquad \text{[A1.23]}$$

A1.3.2. Resolution of a cubic equation – example

We give here a small example to illustrate our purpose. Let us consider the following cubic equation:

$$\alpha^3 - 2\alpha^2 - \alpha + 2 = 0 \qquad \text{[A1.24]}$$

The coefficients (a,b,c,d) are identified from equation [A1.20] as:

$$a = +1, b = -2, c = -1 \text{ and } d = +2 \qquad \text{[A1.25]}$$

We calculate now p and q for determining the nature of the solutions:

$$p = c - \frac{b^2}{3} = -\frac{7}{3} \text{ and } q = d + \frac{2}{27}b^3 - \frac{bc}{3} = \frac{20}{27} \qquad \text{[A1.26]}$$

We calculate the discriminant as:

$$4p^3 + 27q^2 = -36 \leq 0 \qquad \text{[A1.27]}$$

Hence, we have three real solutions for this cubic equation. It can be relevant to compute the following number for the root calculation:

$$Arc\cos\left(\frac{3q}{2p}\sqrt{-\frac{3}{p}}\right) = Arc\cos\left(-\frac{10}{7\sqrt{7}}\right) \approx 2.141173137... \qquad [A1.28]$$

We then compute the three roots of the cubic from equation [A1.23] as:

$$\begin{cases} \alpha_1 = 2\sqrt{\frac{7}{3}}\cos\left[\frac{2.141173137... + 2\pi}{3}\right] + \frac{2}{3} = -1 \\ \alpha_2 = 2\sqrt{\frac{7}{3}}\cos\left[\frac{2.141173137... + 4\pi}{3}\right] + \frac{2}{3} = 1 \\ \alpha_3 = 2\sqrt{\frac{7}{3}}\cos\left[\frac{2.141173137...}{3}\right] + \frac{2}{3} = 2 \end{cases} \qquad [A1.29]$$

Of course, it is easy to check that $\alpha^3 - 2\alpha^2 - \alpha + 2 =$. $(\alpha+1)(\alpha-1)(\alpha-2)$

A1.4. Roots of a quartic function – principle of resolution

We now consider the quartic equation with real coefficients:

$$f(x) = x^4 + ax^3 + bx^2 + cx + d = 0 \text{ with a} \neq 0 \qquad [A1.30]$$

It can be postulated that the quartic corresponds to the beginning of the square of a second-order polynomial equation like:

$$f(x) = \left(x^2 + \frac{a}{2}x + y\right)^2 + \left(-2y - \frac{a^2}{4} + b\right)x^2 + (-ay+c)x+(d-y^2) \qquad [A1.31]$$

where y is a real number. For the following, we will assume that:

$$-\alpha^2 = -2y - \frac{a^2}{4} + b \text{ and } -\beta^2 = d-y^2 \qquad [A1.32]$$

y is chosen such that the second trinome of $f(x)$, constituted of the three last terms of $f(x)$, could be considered in a square format. It is then necessary that:

$$(-ay+c)^2 - 4\left(-2y - \frac{a^2}{4} + b\right)(d-y^2) = 0 \qquad \text{[A1.33]}$$

We recognize a cubic equation expressed with the unknown y:

$$\left(2y - \frac{b}{3}\right)^3 + \left(ac - \frac{b^2}{3} - 4d\right)\left(2y - \frac{b}{3}\right) + \frac{abc}{3} - \frac{2b^3}{27}$$

$$+ \frac{8bd}{3} - a^2d - c^2 = 0 \qquad \text{[A1.34]}$$

which can be solved with the previous Cardano's cubic method. Let y_1, y_2 and y_3 be the three roots of this cubic equation. The parameters α and β will be chosen as:

$$\alpha = \sqrt{2y_1 + \frac{a^2}{4} - b} \quad \text{and} \quad \beta = \sqrt{y_1^2 - d} \quad \text{if } ay_1 - c \geq 0 \text{ and}$$

$$\beta = -\sqrt{y_1^2 - d} \quad \text{if } ay_1 - c < 0 \qquad \text{[A1.35]}$$

Once the cubic root y is calculated $y = y_1$, the quartic function $f(x)$ has the following form:

$$f(x) = \left(x^2 + \frac{a}{2}x + y_1\right)^2 - (\alpha x + \beta)^2$$

$$= \left[x^2 + \left(\frac{a}{2} - \alpha\right)x + y_1 - \beta\right]\left[x^2 + \left(\frac{a}{2} + \alpha\right)x + y_1 + \beta\right] = 0 \qquad \text{[A1.36]}$$

Then, the determination of the roots of the quartic function is reduced to the determination of the roots of two quadratic functions.

Appendix 2

Steel Reinforcement Table

Φ (mm)	Number of bars and steel area (in cm^2)									
	1	2	3	4	5	6	7	8	9	10
5	0.196	0.393	0.589	0.785	0.982	1.178	1.374	1.571	1.767	1.963
6	0.283	0.565	0.848	1.131	1.414	1.696	1.979	2.262	2.545	2.827
8	0.503	1.005	1.508	2.011	2.513	3.016	3.519	4.021	4.524	5.027
10	0.785	1.571	2.356	3.142	3.927	4.712	5.498	6.283	7.069	7.854
12	1.131	2.262	3.393	4.524	5.655	6.786	7.917	9.048	10.179	11.310
14	1.539	3.079	4.618	6.158	7.697	9.236	10.776	12.315	13.854	15.394
16	2.011	4.021	6.032	8.042	10.053	12.064	14.074	16.085	18.096	20.106
20	3.142	6.283	9.425	12.566	15.708	18.850	21.991	25.133	28.274	31.416
25	4.909	9.817	14.726	19.635	24.544	29.452	34.361	39.270	44.179	49.087
32	8.042	16.085	24.127	32.170	40.212	48.255	56.297	64.340	72.382	80.425
40	12.566	25.133	37.699	50.265	62.832	75.398	87.965	100.53	113.10	125.66

Table A2.1. *Abacus of the steel area A_s in cm^2 for each bar diameter Φ (mm)*

Bibliography

[AAS 73] AAS-JAKOBSEN K., "Design of slender reinforced concrete frames", Bericht Nr. 48, Institut für Baustatik, ETH, Zurich, Switzerland, 1973.

[AIF 03] AIFANTIS E.C., "Update on a class of gradient theories", *ACI Structural Journal*, vol. 35, pp. 259–280, 2003.

[AIF 11] AIFANTIS E.C., "On the gradient approach – relation to Eringen's nonlocal theory", *International Journal of Engineering Science*, vol. 49, no. 12, pp. 1367–1377, 2011.

[ALB 81] ALBIGES M., MINGASSON M., *Théorie et pratique du béton armé aux états limites*, Eyrolles, Paris, 1981.

[AME 11] AMERICAN CONCRETE INSTITUTE (ACI Committee 318), Building code requirements for structural concrete (ACI 318-11), and commentary, Farmington Hills, MI, 2011.

[ARI 95] ARISTIZABAL-OCHOA J.D., "Story stability and minimum bracing in RC framed structures: a general approach", *ACI Structural Journal*, vol. 92, no. 6, pp. 735–744, 1995.

[ART 04] ARTEMIADIS N.K., *History of Mathematics: From a Mathematician's Vantage Point*, American Mathematical Society, Paris, 2004.

[AUS 61] AUSTIN W.J., "Strength and design of metal beam columns", *Journal of the Structural Division, ASCE*, vol. 87, no. ST4, pp. 1–32, 1961.

[BAE 08] BAE S., BAYRAK O., "Plastic hinge length of reinforced concrete columns", *ACI Structural Journal*, vol. 105, no. 3, pp. 290–300, 2008.

[BAI 83] BAILOV V., SIGALOV E., *Reinforced Concrete Structures*, MIR, 1983.

[BAK 56] BAKER A.L.L., *Ultimate Load Theory Applied to the Design of Reinforced and Prestressed Concrete Frames*, Concrete Publications Ltd, London, 1956.

[BAR 65] BARNARD P.R., JOHNSON R.P., "Plastic behaviour of continuous composite beams", *Proceedings – Institution of Civil Engineers*, vol. 32, pp. 161–210, 1965.

[BAR 12] BARRERA A.C., BONET J.L., ROMERO M.L., FERNÁNDEZ M.A., "Ductility of slender reinforced concrete columns under monotonic flexure and constant axial load", *Engineering Structures*, vol. 40, pp. 398–412, 2012.

[BAŽ 76] BAŽANT Z.P., "Instability, ductility and size effect in strain-softening concrete", *Journal of Engineering Mechanics, ASCE*, vol. 102, pp. 331–344, 1976.

[BAŽ 87] BAŽANT Z.P., PIJAUDIER-CABOT G., PAN J., "Ductility, snap-back, size effect, and redistribution in softening beams or frames", *Journal of Structural Engineering,* ASCE, vol. 113, no. 12, pp. 2348–2364, 1987.

[BAŽ 03] BAŽANT Z.P., CEDOLIN L., *Stability of Structures – Elastic, Inelastic, Fracture, and Damage Theories*, Dover Publications, Inc., New York, NY, 2003.

[BEN 07] BENALLAL A., MARIGO J.J., "Bifurcation and stability issues in gradient theories with softening", *Modelling and Simulation in Materials Science and Engineering*, vol. 15, pp. 283–295, 2007.

[BIJ 53] BIJLAARD P.P., FISHER P.P., WINTER G., "Strength of columns elastically restrained and eccentrically loaded", *Proceedings,* ASCE, vol. 79, separate no. 292, pp. 1–52, 1953.

[BRI 86] BRIDGE R.Q., FRASER, D.J., "Improved G-factor method for evaluating effective lengths of columns", *Journal of Structural Engineering, ASCE*, vol. 113, no. 6, pp. 1341–1356, 1986.

[BRO 07] BROJAN M., PUKSIC A., KOSEL F., "Buckling and post-buckling of a nonlinearly elastic column", *Zeitschrift für Angewandte Mathematik Mechanik*, vol. 87, no. 7, pp. 518–527, 2007.

[CAL 05] CALGARO J., CORTADE J., *Application de l'Eurocode 2: calcul des bâtiments en béton*, Presses de l'Ecole Nationale des Ponts et Chaussées, 2005.

[CAR 86] CARREIRA D., CHU K.H., "The moment-curvature relationship of RC members", *Advanced Computing: An International Journal,* vol. 83, pp. 191–198, 1986.

[CAS 89] CASANDJIAN C., "Calculs de section en Té en flexion non déviée à l'aide de coefficients adimensionnés", *Rencontres Universitaire de Génie Civil, AUGC* – French Civil Engineering Congress, Rennes, 1989.

[CAS 12] CASANDJIAN C., CHALLAMEL N., LANOS C., HELLESLAND J., *Reinforced concrete beams, columns and frames – Mechanics and Design*, ISTE Ltd, London and John Wiley & Sons, New York, 2012.

[CHA 03] CHALLAMEL N., "A gradient plasticity approach for steel structures", *Comptes-Rendus Mécanique*, vol. 331, no. 9, pp. 647–654, 2003.

[CHA 05a] CHALLAMEL N., LANOS C., CASANDJIAN C., "Creep failure in concrete as a bifurcation phenomenon", *International Journal of Damage Mechanics*, vol. 14, pp. 5–24, 2005.

[CHA 05b] CHALLAMEL N., LANOS C., CASANDJIAN C., "Strain-based anisotropic damage modelling and unilateral effects", *International Journal of Mechanical Sciences*, vol. 47, no. 3, pp. 459–473, 2005.

[CHA 05c] CHALLAMEL N., LANOS C., CASANDJIAN C., "Creep damage modelling for quasi-brittle materials", *European Journal of Mechanics A/Solids*, vol. 24, pp. 593–613, 2005.

[CHA 05d] CHALLAMEL N., HJIAJ M., "Non-local behavior of plastic softening beams", *Acta Mechanica*, vol. 178, pp. 125–146, 2005.

[CHA 06a] CHALLAMEL N., LANOS C., CASANDJIAN C., "Stability analysis of quasi-brittle materials – creep under multiaxial loading", *Mechanics of Time-Dependent Materials*, vol. 10, no. 1, pp. 35–50, 2006.

[CHA 06b] CHALLAMEL N., PIJAUDIER-CABOT G., "Stability and dynamics of a plastic softening oscillator", *International Journal of Solids and Structures*, vol. 43, pp. 5867–5885, 2006.

[CHA 07] CHALLAMEL N., LANOS C., CASANDJIAN C., "Creep failure of a simply-supported beam through a uniaxial continuum damage mechanics model", *Acta Mechanica*, vol. 192, no. 1–4, pp. 213–234, 2007.

[CHA 08a] CHALLAMEL N., WANG C.M., "The small length scale effect for a non-local cantilever beam: a paradox solved", *Nanotechnology*, vol. 19, p. 345703, 2008.

[CHA 08b] CHALLAMEL N., LANOS C., CASANDJIAN C., "Plastic failure of nonlocal beams", *Physics Review E*, vol. 78, p. 026604, 2008.

[CHA 08c] CHALLAMEL N., "A regularization study of some ill-posed gradient plasticity softening beam problems", *Journal of Engineering Mathematics*, vol. 62, pp. 373–387, 2008.

[CHA 09a] CHALLAMEL N., "An application of large displacement limit analysis to frame structures", *Structural Engineering & Mechanics*, vol. 33, no. 2, pp. 169–177, 2009.

[CHA 09b] CHALLAMEL N., LANOS C., CASANDJIAN C., "Some closed-form solutions to simple beam problems using non-local (gradient) damage theory", *International Journal of Damage Mechanics*, vol. 18, no. 6, pp. 569–598, 2009.

[CHA 09c] CHALLAMEL N., LANOS C., CASANDJIAN C., "Comment une poutre peut-elle casser?", *Annales du Bâtiment et des Travaux Publics*, vol. 6, pp. 42–46, 2009.

[CHA 09d] CHALLAMEL N., RAKOTAMANANA L., LE Marrec L., "A dispersive wave equation using non-local elasticity", *Comptes-Rendus Mécanique*, vol. 337, pp. 591–595, 2009.

[CHA 09e] CHANDRASEKARAN S., NUNZIANTE L., SERINO G., CARANNANTE F., *Seismic Design Aids for Nonlinear Analysis of Reinforced Concrete Structures*, CRC Press/Taylor and Francis, Florida, FL, 2009.

[CHA 10a] CHALLAMEL N., "A variationally-based nonlocal damage model to predict diffuse microcracking evolution", *International Journal of Mechanical Sciences*, vol. 52, pp. 1783–1800, 2010.

[CHA 10b] CHALLAMEL N., LANOS C., CASANDJIAN C., "On the propagation of localization in the plasticity collapse of hardening-softening beams", *International Journal of Engineering Science*, vol. 48, no. 5, pp. 487–506, 2010.

[CHA 10c] CHALLAMEL N., MEFTAH S.A., BERNARD F., "Buckling of elastic beams on nonlocal foundation: a revisiting of Reissner model", *Mechanics Research Communications*, vol. 37, pp. 472–475, 2010.

[CHA 10d] CHANDRASEKARAN S., NUNZIANTE L., SERINO G., CARANNANTE F., "Axial force bending moment limit domain and flow rule for reinforced concrete elements using Eurocode", *International Journal of Damage Mechanics*, vol. 19, no. 5, pp. 523–558, 2010.

[CHA 11a] CHALLAMEL N., GIRHAMMAR U.A., "Boundary layer effect in composite beams with interlayer slip", *Journal of Aerospace Engineering, ASCE*, vol. 24, no. 2, pp. 199–209, 2011.

[CHA 11b] CHALLAMEL N., GIRHAMMAR U.A., "Variationally-based theories for buckling of partially composite beam-columns including shear and axial effects", *Engineering Structures*, vol. 33, no. 8, pp. 2297–2319, 2011.

[CHA 11c] CHALLAMEL N., HELLESLAND J., "Simplified buckling analysis of imperfection sensitive reinforced concrete columns", *Proceedings of the 24th Nordic Seminar on Computational Mechanics*, 3–4 November 2011, Aalto University, Helsinki, Finland.

[CHA 11d] CHANDRASEKARAN S., NUNZIANTE L., SERINO G., CARANNANTE F., "Curvature ductility of RC sections based on Eurocode: analytical procedure", *KSCE Journal of Civil Engineering*, vol. 15, no. 1, pp. 131–144, 2011.

[CHA 12a] CHALLAMEL N., HELLESLAND J., "Buckling of imperfect CDM structural systems", *International Conference on Damage Mechanics (ICDM)*, Belgrade, Serbia, 25–27 June 2012.

[CHE 99] CHEONG-SIAT-MOY F., "An improved K-factor formula", *Journal of Structural Engineering*, vol. 125, no. 2, pp. 169–174, 1999.

[CIM 07] CIMETIERE A., HALM D., MOLINES E., "A damage model for concrete beam in compression", *Mechanics Research Communications*, vol. 34, pp. 91–96, 2007.

[COM 77] COMITÉ EURO-INTERNATIONAL DU BETON (CEB), CEB/FIP design manual on buckling and instability", CEB Bulletin d'Information, vol. 123, 1977.

[COM 78] COMITÉ EURO-INTERNATIONAL DU BETON (CEB), CEB-FIP model code for concrete structures, CEB Bulletin d'Information, vol. 124–125, 1978.

[COM 93] COMITÉ EURO-INTERNATIONAL DU BETON (CEB), CEB-FIP model code 1990, CEB Bull., vol. 213–214, 1993.

[DAN 08] DANIELL J.E., OEHLERS D.J., GRIFFITH M.C., MOHAMED ALI M.S., OZBAKKALOGLU T., "The softening rotation of reinforced concrete members", *Engineering Structures*, vol. 30, pp. 3159–3166, 2008.

[DEB 92] DE BORST R., MÜHLHAUS H.B., "Gradient-dependent plasticity: formulation and algorithmic aspects", *International Journal for Numerical Methods in Engineering*, vol. 35, pp. 521–539, 1992.

[DES 64] DESAYI P., KRISHNAN S., "Equations for the stress-strain curve of concrete", *Advanced Computing: Journal of American Concrete Institute*, vol. 61, no. 3, pp. 345–350, 1964.

[DES 05] Design aids for EC2, Design of concrete structures, Design aids for ENV 1992-1-1 Eurocode 2, Part1, Betonvereniging, The Concrete Society, E & FN Spon, 2005.

[DI 09] DI PAOLA M., MARINO F., ZINGALES M., "A generalized model of elastic foundation based on long-range interactions: integral and fractional model", *International Journal of Solids and Structures*, vol. 46, no. 17, pp. 3124–3137, 2009.

[ELI 12] ELISHAKOFF I., PENTARAS D., DUJAT K., VERSACI C., MUSCOLINO G., STORCH J., BUCAS S., CHALLAMEL N., NATSUKI T., ZHANG Y.Y., WANG C.M., GHYSELINCK G., *Carbon Nanotubes and Nanosensors: Vibrations, Buckling and Ballistic Impact*, ISTE Ltd, London and John Wiley & Sons, New York, 2012.

[ENG 03] ENGELEN R.A.B., GEERS M.G.D., BAAIJENS F.P.T., "Nonlocal implicit gradient-enhanced elasto-plasticity for the modelling of softening behaviour", *International Journal of Plasticity*, vol. 19, pp. 403–433, 2003.

[ERI 83] ERINGEN A.C., "On differential equations of nonlocal elasticity and solutions of screw dislocation and surface waves", *Journal of Applied Physics*, vol. 54, pp. 4703–4710, 1983.

[EUR 04] EUROPEAN COMMITTEE FOR STANDARDIZATION (CEN), EN 1992-1-1:2004:E– Eurocode 2: design of concrete structures – Part 1-1: general rules and rules for buildings, Brussels, Belgium, 2004.

[EUR 08a] Eurocode 2 Worked Examples, European Concrete Platform ASBL, May 2008.

[EUR 08b] EUROPEAN COMMITTEE FOR STANDARDISATION (CEN)/Standard Norge, Eurocode 2: Design of concrete structures -- Part 1-1: General rules and rules for buildings (NS-EN 1992-1-1:2004 + National Annex NA:2008), Oslo, Norway, 2008.

[FAE 73] FAESSEL P., MORISSET A., FOURE B., "Le flambement des poteaux en béton armé", *Annales de l'ITBTP*, p. 305, 1973.

[FOR 09] FOREST S., "Micromorphic approach for gradient elasticity, viscoplasticity, and damage", *Journal of Engineering Mechanics*, vol. 135, no. 3, pp. 117–131, 2009.

[FRE 96] FRÉMOND M., NEDJAR B., "Damage, gradient of damage and principle of virtual power", *International Journal of Solids and Structures*, vol. 33, pp. 1083–1103, 1996.

[FRI 94] FRISLID A., Evaluation of simplified methods for stability analysis of frame structures, Candidatus Scientiarum thesis, Mechanics Division, Department of Mathematics, University of Oslo, Norway, 1994.

[FUE 78] FUENTES A., *Béton armé – Calcul des ossatures – Torsion – Flambement – Oscillations – Déformations plastiques*, Eyrolles, 1978.

[GAL 68] GALAMBOS T.V., *Structural Members and Frames*, Prentice-Hall, Inc., New York, NY, 1968.

[GAL 02] GALILEO, "Discorsi e Dimonstrazioni Matematiche, intorno à due nuove Scienze, 1638", in HAWKINGS S. (ed.), *Sur les épaules des géants – les plus grands textes de physique et d'astronomie*, Dunod, pp. 154–182, 2002.

[GAL 08] GALAMBOS, T.V., SUROVEK, A.E., *Structural Stability of Steel – Concepts and Applications for Structural Engineers*, John Wiley & Sons, Inc., Hoboken, NJ, 2008.

[GIR 07] GIRHAMMAR U.A., PAN D.H., "Exact static analysis of partially composite beams and beam-columns", *International Journal of Mechanical Sciences*, vol. 49, pp. 239–255, 2007.

[GYÖ 88] GYÖRGY F., *Cours de béton armé et constructions hydrauliques, Tome 1 – Béton armé aux états limites*, Presse de l'Ecole Nationale Polytechnique d'Alger, 1988.

[HAL 96] HALM D., DRAGON A., "A model of anisotropic damage by mesocrack growth: unilateral effect", *International Journal of Damage Mechanics*, vol. 5, pp. 384–402, 1996.

[HAS 85] HASLACH H.W., "Post-buckling behaviour of columns with nonlinear constitutive equations", *International Journal of Non-Linear Mechanics*, vol. 20, no. 1, pp. 53–67, 1985.

[HAS 09] HASKETT M., OEHLERS D.J., ALI M.S.M., WU C., "Rigid moment-rotation mechanism for reinforced concrete beam hinges", *Engineering Structures*, vol. 31, pp. 1032–1041, 2009.

[HEL 70a] HELLESLAND J., A study into the sustained and cyclic load behaviour of reinforced concrete columns, PhD Thesis, Department of Civil Engineering, University of Waterloo, Canada, 1970.

[HEL 70b] HELLESLAND J., GREEN R., "Strength characteristics of reinforced concrete columns under sustained loading", *IABSE Symposium on Design of Concrete Structures for Creep, Shrinkage and Temperature Changes*, Madrid, Spain, 1970.

[HEL 72] HELLESLAND J., GREEN R., "A stress and time dependent strength law for concrete", *Cement and Concrete Research*, vol. 2, pp. 261–275, 1972.

[HEL 76] HELLESLAND J., Approximate second order analysis of unbraced frames, Technical report, Dr. Ing. A. Aas-Jakobsen Ltd., Oslo, Norway, pp. 1–43, 1976.

[HEL 81] HELLESLAND J., SCORDELIS A.C., "Analysis of r.c. bridge columns under imposed deformations", *Proceedings, IABSE Colloquium on Advanced Mechanics of Reinforced Concrete*, Delft, Holland, pp. 545–559, June, 1981.

[HEL 85] HELLESLAND, J., CHOUDHURY, D., SCORDELIS, A.C., Nonlinear analysis and design of RC bridge columns subjected to imposed deformations, Report no. UCB/SESM-85/03, Department of Civil Engineering, University of California, Berkeley, CA, 1985.

[HEL 90a] HELLESLAND J., "Ny NS~3473 – Når er en trykkstav slank?", Betongprodukter, *Norges Betongindustriforbund*, vol. 1, 1990.

[HEL 90b] HELLESLAND J., "Øvre slankhetsgrenser for trykkstaver", Betongprodukter, *Norges Betongindustriforbund*, vol. 4, 1990.

[HEL 95] HELLESLAND J., "Simplified system instability analysis", in SHANMUGAN N.E., CHOO Y.S. (eds.), *4th Pacific Structural Steel Conference (PSSC'95)*, vol. 1, Pergamon Press, pp. 95–102, 1995.

[HEL 96a] HELLESLAND J., BJORHOVDE R., "Restraint demand factors and effective lengths of braced columns", *Journal of Structural Engineering, ASCE*, vol. 122, no. 10, pp. 1216–1224, 1996.

[HEL 96b] HELLESLAND J., BJORHOVDE R., "Improved frame stability analysis with effective lengths", *Journal of Structural Engineering, ASCE*, vol. 122, no. 11, pp. 1275–1283, 1996.

[HEL 97] HELLESLAND, J., FRISLID, A., "Approximate critical load analysis of frame systems with axially compressed beams", in USAMI T. (ed.), *Proceedings, 5th International Colloquium and Ductility of Structures (SDSS'97)*, Japanese Society of Steel Construction, Nagoya University, Japan, vol. 2, pp. 699–706, 1997.

[HEL 98] HELLESLAND J., "Application of the method of means to the stability analysis of unbraced frames", *Journal of Constructional Steel Research*, vol. 46, no. 1–3, 1998. (The full version is available in Preprint Series for Mechanics and Applied Mathematics, no. 4, December 1997, Department of Mathematics, University of Oslo, Norway).

[HEL 02a] HELLESLAND J., Lower slenderness limits for braced end-loaded r.c. compression members, Research report in mechanics, no. 02-2, Mechanics Division, University of Oslo, Oslo, Norway, pp. 1–33, 2002.

[HEL 02b] HELLESLAND J., Lower slenderness limits for unbraced and transversely loaded r.c. compression members, Research report in mechanics, no. 02-1, Mechanics Division, University of Oslo, Oslo, Norway, pp. 1–37, 2002.

[HEL 05a] HELLESLAND J., "Nonslender column limits for braced and unbraced reinforced concrete members", *ACI Structural Journal*, vol. 102, no. 1, pp. 12–21, 2005.

[HEL 05b] HELLESLAND J., Analysis of second-order effects with axial loads, EN 1992-1-1:2004, National Annex, Evaluations and proposals submitted to Standard Norge, Oslo, Norway, pp. 1–10, 2005.

[HEL 07] HELLESLAND J., "Mechanics and effective lengths of columns with positive and negative end restraints", *Engineering Structures*, vol. 29, no. 12, pp. 3464–3474, 2007.

[HEL 08a] HELLESLAND J., Approximate second order analysis of unbraced frames reflecting inter-storey interaction in single curvature regions. Research report in mechanics, no. 08-2. Mechanics Division, University of Oslo, Oslo, Norway, pp. 1–26, 2008.

[HEL 08b] HELLESLAND J., "Mechanics and slederness limits of sway-restricted reinforced concrete columns", *Journal of Structural Engineering, ASCE*, vol. 134, no. 8, pp. 1300–1309, 2008.

[HEL 09a] HELLESLAND J., "Extended second order approximate analysis of frames with sway-braced column interaction", *Journal of Constructional Steel Research*, vol. 65, no. 5, pp. 1075–1086, 2009.

[HEL 09b] HELLESLAND J., "Second order approximate analysis of unbraced multistorey frames with single curvature regions", *Engineering Structures*, vol. 31, no. 8, pp. 1734–1744, 2009.

[HEL 12] HELLESLAND J., "Evaluation of effective length formulas and applications in system instability analysis", *Engineering Structures*, vol. 45, no. 12, pp. 405–420, 2012.

[HEY 99] HEYMAN J., *The Science of Structural Engineering*, Imperial College Press, 1999.

[HIL 58] HILL R., "A general theory of uniqueness and stability in elastic-plastic solids", *Journal of the Mechanics and Physics of Solids*, vol. 6, pp. 236–249, 1958.

[HIL 76] HILLERBORG A., MODEER M., PETERSSON P.E., "Analysis of crack formation and crack growth in concrete by means of fracture mechanics and finite elements", *Cement and Concrete Research*, vol. 6, pp. 773–782, 1976.

[HOG 51] HOGNESTAD E., A study of combined and axial load in reinforced concrete members, Bulletin no. 399, University of Illinois Engineering Experiment Station, Urbana, November 1951.

[HOR 75] HORNE M.R., "An approximate method for calculating the elastic critical loads of multi-storey plane frames", *Journal of Structural Engineering*, vol. 53, no. 6, pp. 242–248, 1975.

[JIR 02] JIRÁSEK M., BAŽANT Z.P., *Inelastic Analysis of Structures*, Wiley, 2002.

[JUN 05] JUNG J.H., KANG T.J., "Large deflection analysis of fibers with nonlinear elastic properties", *Journal of the Textile Institute*, vol. 75, no. 10, pp. 715–723, 2005.

[KAB 66] KABAILA A.P., HALL A.S., "Analysis of instability of unrestrained prestressed concrete columns with end eccentricities", in *Symposium on Reinforced Concrete Columns*, presented at the 61th Annual ACI Convention, San Francisco, California, March 4, 1965, Compiled under the sponsorship of ACI-ASCE Committee 441, American Concrete Institute, Publication SP-13, 1966.

[KOE 07] KOECHLIN P., POTAPOV S., "Global constitutive model for reinforced concrete plates", *Journal of Engineering Mechanics*, vol. 133, no. 3, pp. 257–266, 2007.

[KOE 08] KOECHLIN P., ANDRIEUX S., MILLARD A., POTAPOV S., "Failure criterion for reinforced concrete beams and plates subjected to membrane force, bending and shear", *European Journal of Mechanics A/Solids*, vol. 27, pp. 1161–1183, 2008.

[KOI 45] KOITER W.T., Over de stabiliteit van het elastische evenwicht, Dissertation, Delft, Holland, The Netherlands, 1945.

[KOU 87] KOUNADIS A.N., MALLIS J.G., "Elastica type buckling analysis of bars from nonlinearly elastic material", *International Journal of Non-Linear Mechanics*, vol. 22, no. 2, pp. 99–107, 1987.

[KRA 11] KRAUBERGER N., BRATINA S., SAJE M., SCHNABL S., PLANINC I., "Inelastic buckling load of a locally weakened reinforced concrete column", *Engineering Structures*, vol. 34, pp. 278–288, 2011.

[KUH 76] KUHN G.K., An appraisal of the effective length alignment charts, Engineering report for the degree of M.Sc. in engineering, Arizona State University, Tempe, AZ 1976.

[LAC 08] LACARBONARA W., "Buckling and post-buckling of non-uniform non-linearly elastic rods", *International Journal of Mechanical Sciences*, vol. 50, pp. 1316–1325, 2008.

[LAI 83a] LAI S.M.A., MACGREGOR J.G., "Geometric non-linearities in unbraced multistory frames", *Journal of Structural Engineering, ASCE*, vol. 109, no. 11, pp. 2528–2545, 1983.

[LAI 83b] LAI S.M.A., MACGREGOR J.G., HELLESLAND J., "Geometric non-linearities in nonsway frames", *Journal of Structural Engineering, ASCE*, vol. 109, no. 12, pp. 2770–2785, 1983.

[LAR 12] LARSEN K.P., POULSEN P.N., NIELSEN L.O., "Limit analysis of 3D reinforced concrete beam elements", *Journal of Engineering Mechanics*, vol. 138, no. 3, pp. 286–296, 2012.

[LEE 49] LEE A.Y.-W., A study on column analysis, PhD Thesis, Cornell University, Ithaca, NY, 1949.

[LEE 09] LEE C.L., FILIPPOU F.C., "Efficient beam-column element with variable inelastic end zones", *Journal of Structural Engineering*, vol. 135, no. 11, pp. 1310–1319, 2009.

[LEM 05] LEMAITRE J., DESMORAT R., *Engineering Damage Mechanics: Ductile, Creep, Fatigue and Brittle Failures*, Springer, 2005.

[LEM 77] LEMESSURIER W.M., "A practical method of second order analysis", *Engineering Journal, AISC*, vol. 14, no. 2, pp. 49–67, 1977.

[LEN 81] LENSCHOW R., *Betongkonstruksjoner*, Tapir, 1981.

[LEO 78] LEONHARDT F., *Vorlesungen über Massivbau; Vierter Teil, Nachweis der Gebrauchsfähigskeit; Rissebeschränkung, Formänderungen, Momentenumlagerung und Bruchlinientheorie im Stahlbetonbau*, Springer-Verlag, 1978.

[LEW 87] LEWIS G., MAZILU P., MONASA F., "A variational approach for the deflections and stability behaviour of postbuckled elastic-plastic slender struts", *International Journal of Non-Linear Mechanics*, vol. 22, no. 5, pp. 373–385, 1987.

[LHE 76] L'HERMITE R., *Flambage et stabilité – Le flambage élasto-plastique des colonnes et systèmes de barres droites*, Eyrolles, 1976.

[LOR 99] LORENTZ E., ANDRIEUX S., "A variational formulation for nonlocal damage models", *International Journal of Plasticity*, vol. 15, pp. 119–138, 1999.

[LUI 92] LUI E.M., "A novel approach for K factor determination", *Engineering Journal, AISC*, vol. 29, no. 4, pp. 150–159, 1992.

[MAC 75] MACGREGOR J.G., OELHAFEN V.H., HAGE S.E., "A re-examination of the EI values for slender columns", *Reinforced Concrete Columns*, Special Publication SP-50, American Concrete Institute, pp. 1–40, 1975.

[MAC 97] MAC GREGOR J.G., *Reinforced Concrete Mechanics and Design*, 3rd ed., Prentice Hall, 1997.

[MAG 14] MAGNY M.A.V., *La construction en béton armé, théorie et pratique*, ED Librairie Polytechnique, CH. Béranger, 1914.

[MAN 67] MANUEL R.F., MACGREGOR J.G., "Analysis of restrained reinforced concrete columns under sustained load", *Advanced Computing: A Journal of American Concrete Institute*, vol. 64, no. 1, pp. 12–23, 1967.

[MAR 03] MARI A., HELLESLAND J., Lower slenderness limits for reinforced concrete columns, Research Report no. 2003-01, Department of Construction Engineering, Universitat Politecnica de Catalunya, Barcelona, Spain, pp. 1–50, 2003.

[MAR 05] MARI A.R., HELLESLAND J., "Lower slenderness limits for rectangular reinforced concrete columns", *Journal of Structural Engineering*, vol. 131, no. 1, pp. 85–95, 2005.

[MAS 59] MASSONNET C., "Stability considerations in the design of steel columns", *Journal of the Structural Division, ASCE*, vol. 85, pp. 75–111, 1959.

[MAT 67] MATTOCK A.H., "Discussion of "rotational capacity of reinforced concrete beams" by W.G. Corley", *Journal of the Structural Division, ASCE*, vol. 93, no. ST2, pp. 519–522, 1967.

[MAZ 86] MAZARS J., "A description of micro and macroscale damage of concrete structures", *Engineering Fracture Mechanics*, vol. 25, no. 5–6, pp. 729–737, 1986.

[MAZ 96] MAZARS J., PIJAUDIER-CABOT G., "From damage to fracture mechanics and conversely: a combined approach", *International Journal of Solids and Structures*, vol. 33, pp. 3327–3342, 1996.

[MAZ 09] MAZARS J., MILLARD A. (eds), *Dynamic Behaviour of Concrete and Seismic Engineering*, ISTE Ltd, London and John Wiley & Sons, New York 2009.

[MEN 01] MENDIS P., "Plastic hinge lengths of normal and high-strength concrete in flexure", *Advances in Structural Engineering*, vol. 4, pp. 189–195, 2001.

[MON 74] MONASA F.E., "Deflections and stability behavior of elastoplastic flexible bars", *Journal of Applied Mechanics*, vol. 96, pp. 537–538, 1974.

[MOS 07] MOSLEY B., BUNGAY J., HULSE R., *Reinforced Concrete Design to Eurocode 2*, Palgrave MacMillan, 2007.

[MUH 91] MÜHLHAUS H.B., AIFANTIS E.C., "A variational principle for gradient plasticity", *International Journal of Solids and Structures*, vol. 28, pp. 845–857, 1991.

[MUR 97] MURAKAMI S., KAMIYA K., "Constitutive and damage evolution equations of elastic-brittle materials based on irreversible thermodynamics", *International Journal of Mechanical Sciences*, vol. 39, no. 4, pp. 473–486, 1997.

[MUR 12] MURAKAMI S., *Continuum Damage Mechanics – A Continuum Mechanics Approach to the Analysis of Damage and Fracture*, Springer, 2012.

[NGU 11a] NGUYEN Q.S., "Variational principles in the theory of gradient plasticity", *Comptes-Rendus Mécanique*, vol. 339, pp. 743–750, 2011.

[NGU 11b] NGUYEN V.P., LLOBERAS-VALLS O., STROEVEN M., SLUYS L.J., "Homogenization-based multiscale crack modelling: from micro-diffusive damage to macro-cracks", *Computer Methods in Applied Mechanics and Engineering*, vol. 200, pp. 1220–1236, 2011.

[NIE 99] NIELSEN M.P., *Limit Analysis and Concrete Plasticity*, CRC, Boca Raton, 1999.

[NOR 89] NORWEGIAN STANDARDS ASSOCIATION (NSF), NS 3473—Concrete Structures. Design Rules, 3rd ed., Oslo, Norway (Standard Norge), 1989; 5th ed., 1998.

[NOR 04] NORME EUROPÉENNE EN 1992-1.1 EUROCODE 2 – Calcul des structures en béton, AFNOR, April 2004.

[ODE 70] ODEN J.T., CHILDS S.B., "Finite deflection of a nonlinearly elastic bar", *Journal of Applied Mechanics*, vol. 69, pp. 48–52, 1970.

[OTT 05] OTTOSEN N.S., RISTINMAA M., *The Mechanics of Constitutive Modelling*, Elsevier, 2005.

[PAI 09] PAILLÉ J.M., *Calcul des structures en béton*, Guide d'application, Eyrolles, 2009.

[PAR 75] PARK R., PAULAY T., *Reinforced Concrete Structures*, John Wiley & Sons, New York, NY, 1975.

[PAU 92] PAULAY T., PRIESTLEY M.J.N., *Seismic Design of Reinforced Concrete and Masonry Buildings*, Wiley, New York, NY, 1992.

[PAU 11] PAULTRE P., *Structures en béton armé – Analyse et dimensionnement*, International Polytechnic Press, Polytechnic School of Montreal , 2011.

[PEE 96] PEERLINGS R.H.J., DE BORST R., BREKELMANS W.A.M., DE VREE J.H.P., "Gradient-enhanced damage for quasi-brittle materials", *International Journal for Numerical Methods in Engineering*, vol. 39, pp. 3391–3403, 1996.

[PER 09] PERCHAT J., *Eurocode 2 – Béton armé, Dispositions et données générales*, Techniques de l'Ingénieur, 2009.

[PER 10] PERCHAT J., *Traité de béton armé*, Editions Le Moniteur, 2010.

[PFR 64] PFRANG E.O., SIESS C., SOZEN M.A., "Load-moment-curvature characteristics of RC cross sections", *Advanced Computing: An International Journal*, vol. 61, no. 7, pp. 763–778, 1964.

[PHA 10a] PHAM K., MARIGO J.J., "The variational approach to damage: I. The foundations", *Comptes-Rendus Mécanique*, vol. 338, pp. 191–198, 2010.

[PHA 10b] PHAM K., MARIGO J.J., "The variational approach to damage: II. The gradient damage models", *Comptes-Rendus Mécanique*, vol. 338, pp. 199–206, 2010.

[PIJ 87] PIJAUDIER-CABOT G., BAŽANT Z.P., "Nonlocal damage theory", *Journal of Engineering Mechanics*, vol. 113, pp. 1512–1533, 1987.

[PLA 93] PLANAS J., ELICES M., GUINEA G.V., "Cohesive cracks versus nonlocal models: closing the gap", *International Journal of Fracture*, vol. 63, pp. 173–187, 1993.

[POL 98] POLIZZOTTO C., BORINO G., FUSCHI P., "A thermodynamically consistent formulation of nonlocal and gradient plasticity", *Mechanics Research Communications*, vol. 25, no. 1, pp. 75–82, 1998.

[PRI 87] PRIESTLEY M.J.N., PARK R., "Strength and ductility of concrete bridge columns under seismic loading", *ACI Structural Journal*, vol. 84, no. 1, pp. 61–76, 1987.

[REI 58] REISSNER E., "A note on deflections of plates on a viscoelastic foundation", *Journal of Applied Mechanics*, vol. 25, pp. 144–145, 1958.

[ROB 74] ROBINSON J.R., *Cours de béton armé de l'Ecole Nationale des Ponts et Chaussées*, 1974.

[ROB 75] ROBINSON J., FOURÉ F., BOURGHLI M., "Le flambement des poteaux en béton armé chargés avec des excentricités différentes à leurs extrémités", *Annales de l'ITBTP*, vol. 333, pp. 46–74, 1975.

[ROU 09a] ROUX J., *Pratique de l'Eurocode 2 (Tome 1)*, Eyrolles, 2009.

[ROU 09b] ROUX J., *Pratique de l'Eurocode 2 (Tome 2)*, Eyrolles, 2009.

[ROY 01] ROYER-CARFAGNI G., "Can a moment-curvature relationship describe the flexion of softening beams", *European Journal of Mechanics A/Solids*, vol. 20, pp. 253–276, 2001.

[RUB 73] RUBIN H., "Das QΔ Verfahren zur vereinfachten Berechnung verschieblicher Rahmensysteme nach dem Traglastverfahren der Theorie II. Ordnung", *Der Bauingenieur*, vol. 48, no. 8, pp. 275–285, 1973.

[RUS 60] RÜSCH H., "Researches toward a general flexural theory for structural concrete", *Advanced Computing: A Journal of American concrete Institute*, vol. 57, no. 1, pp. 1–28, 1960.

[SAL 83] SALENÇON J., *Calcul à la rupture et analyse limite*, Presses de l'École Nationale des Ponts et Chaussées, 1983.

[SAL 90] SALENÇON J., "An introduction to the yield design theory and its application to soil mechanics", *European Journal of Mechanics A/Solids*, vol. 9, no. 5, pp. 477–500, 1990.

[SAL 02] SALENÇON J., *De l'élastoplasticité au calcul à la rupture*, Editions de l'Ecole Polytechnique, Ellipses, 2002.

[SAL 06] SALENÇON J., "Revisiting Galileo's insight in structural mechanics", *Conference at the Department of Structural Mechanics*, Budapest University of Technology and Economics, June 2006.

[SAR 68] SARGIN M., Stress-strain relationships for concrete and the analysis of structural concrete sections, PhD Thesis, University of Waterloo, Ontario, Canada, p. 334, submitted March 1968.

[SAR 69] SARGIN M., HANDA V.K., A General Formulation for the Stress-Strain Properties of Concrete, SM Report Solid Mechanics Division, University of Waterloo, Canada, No. 3, May 1969.

[SAR 71] SARGIN M., *Stress-Strain Relationships for Concrete and the Analysis of Structural Concrete Sections*, Solid Mechanics Division, University of Waterloo, Ontario, Canada, 1971.

[SAW 65] SAWYER H.A., "Design of concrete frames for two failure states", *Proceedings of the International Symposium on the Flexural Mechanics of Reinforced Concrete*, ASCE-ACI, Miami, pp. 405–431, 1965.

[SIE 10] SIEFFERT Y., *Le béton armé selon les Eurocodes 2*, Dunod, 2010.

[THO 09] THONIER H., *Conception et calcul de structures de bâtiments – l'Eurocode 2 pratique*, Presses de l'Ecole Nationale des Ponts et Chaussées, 2009.

[TIM 61] TIMOSHENKO S.P., GERE J.M., *Theory of Elastic Stability*, McGraw-Hill, 1961.

[TIM 83] TIMOSHENKO S.P., *History of Strength of Materials*, Dover Publications, 1983.

[VER 94] VERMEER P.A., BRINKGREVE R.B.J., "A new effective non-local strain measure for softening plasticity", in CHAMBON, R., DESRUES, J., VARDOULAKIS, I. (eds), *Localisation and Bifurcation Theory for Soils and Rocks*, Balkema, Rotterdam, pp. 89–100, 1994.

[VIR 04] VIRGIN L.N., PLAUT, R.H., "Postbuckling and vibration of linearly elastic and softening columns under self-weight", *International Journal of Solids and Structures*, vol. 41, pp. 4989–5001, 2004.

[WAL 90] WALTHER R., MIEHLBRADT M., *Dimensionnement des structures en béton, Bases et Technologie*, Treaty of Civil Engineering at the Ecole Polytechnique Fédérale de Lausanne, vol. 7, 1990.

[WAN 96] WANG C.Y., "Global buckling load of a nonlinearly elastic bar", *Acta Mechanica*, vol. 119, pp. 229–234, 1996.

[WES 04] WESTERBERG B., Second-order effects in slender concrete structure – background to rules EC2, TRITA-BKN, Report 77, Betonbyggnad, KTH, Stockholm, Sweden, p. 98, 2004.

[WIN 54] WINTER G., "Compression members in trusses and frames", *The Philosophy of Column Design, Proceedings of the 4th Technical Session*, Column Research Council, Lehigh University, Bethlehem, PA, 1954.

[WOO 68] WOOD R.H., "Some controversial and curious developments in the plastic theory of structures", in HEYMAN J., LECKIE F.A. (eds), *Engineering Plasticity*, Cambridge University Press, UK, pp. 665–691, 1968.

[XU 02] XU L., LIU, Y., "Story stability of semi-braced steel frames", *Journal of Constructional Steel Research*, vol. 58, no. 4, pp. 467–491, 2002.

[YU 82] YU T.X., JOHNSON W., "The plastica: the large elastic–plastic deflection of a strut", *International Journal of Non-Linear Mechanics*, vol. 17, no. 3, pp. 195–209, 1982.

[YU 96] YU T.X., ZHANG L.C., *Plastic Bending: Theory and Applications*, World Scientific, Singapore, 1996.

[YU 12] YU Y., WU B., LIM C.W., "Numerical and analytical approximations to large post-buckling deformation of MEMS", *International Journal of Mechanical Sciences*, vol. 55, pp. 95–103, 2012.

[ZHA 10] ZHANG Y.Y., WANG C.M., CHALLAMEL N., "Bending, buckling and vibration of hybrid nonlocal beams", *Journal of Engineering Mechanics*, vol. 136, no. 5, pp. 562–574, 2010.

[ZHA 12] ZHAO X.M., WU Y.F., LEUNG A.Y.T., "Analyses of plastic hinge regions in reinforced concrete beams under monotonic loading", *Engineering Structures*, vol. 34, pp. 466–482, 2012.

Index